ENGINEERING ETHICS

DEBORAH G. JOHNSON

ENGINEERING ETHICS

Contemporary and Enduring Debates

Yale

UNIVERSITY

PRESS

NEW HAVEN AND LONDON

Published with assistance from the Mary Cady Tew Memorial Fund.

Yale University Press books may be purchased in quantity for educational, business, or promotional use. For information, please e-mail sales.press@yale.edu (US office) or sales@yaleup.co.uk (UK office).

Set in New Aster type by IDS Infotech Ltd., Chandigarh, India.
Printed in the United States of America.

Library of Congress Control Number: 2019948567
ISBN 978-0-300-20924-2 (pbk. : alk. paper)

A catalogue record for this book is available from the British Library.

This paper meets the requirements of ANSI/NISO Z39.48–1992 (Permanence of Paper).

10 9 8 7 6 5 4 3 2 1

CONTENTS

Acknowledgments   *vii*

Introduction   1

PART ONE **Foundational Issues**

1 Can Engineering Ethics Be Taught?   9

2 Do Engineers Need Codes of Ethics?   26

3 How Should Engineers Think about Ethics?   46

PART TWO **Employment Relationships**

4 Should Engineers See Themselves as Guns
       for Hire?   73

5 Are Whistleblowing Engineers Heroes or Traitors?   93

PART THREE **Engineers, Safety, and Social Responsibility**

6 Are Rotten Apples or Rotten Barrels Responsible
      for Technological Mishaps?   117

7 Will Autonomous Cars Ever Be Safe Enough?   137

8 Is Social Justice in the Scope of Engineers' Social
      Responsibilities?   156

      Conclusion   177

      Notes   181

      Index   191

ACKNOWLEDGMENTS

I am grateful to the following colleagues and students who provided invaluable comments on chapters at various stages of development: Rich Burgess, Missy Cummings, Rider Foley, Joe Herkert, Lauren Purnell, Jacob Rogerson, Claire Trevisan, Shannon Vallor, and Mario Verdicchio. The book is much better because of their input. I am also thankful for the useful feedback on an early draft of the entire manuscript from the anonymous reviewers for Yale University Press. While writing this book, I was partially funded by the Norwegian Research Council as part of the ViSmedia Project in the Department of Information Science and Media Studies at the University of Bergen.

ENGINEERING ETHICS

INTRODUCTION

THROUGHOUT human history, technology has shaped what people do as well as how they do it and has affected the quality and character of human lives. Awareness of technology's powerful role in constituting human societies intensified in the late twentieth century and continues to grow in the twenty-first century. Perhaps because of the dropping of the first atomic bomb, and certainly because of such varied inventions as the automobile, nuclear power, birth control pills, the Internet, and cell phones, we now understand that technology is a major force that influences the character of human relationships, social arrangements, and institutional organization and that it will drive the future.

The role of engineers and engineering in all of this is complicated because although inventors and entrepreneurs have always played a role in the long history of technology, engineering as a profession did not develop until the late nineteenth century. In the United States, for example, the American Society of Civil Engineers formed in 1852, the American Institute of Mining and Metallurgical Engineers was founded in 1871, and the American Society of Mechanical Engineers was organized in 1880. Licensure in engineering began only a hundred years ago.[1]

Interest in the ethical issues associated with technology and engineering is arguably even younger and appears not to have arisen, at least formally, until the mid-twentieth century. To my knowledge, the first American book whose title refers to ethics and engineering is *Ethical Problems in Engineering*, published in 1965.[2] Formal courses on engineering ethics at American universities were not taught until the late 1970s and early 1980s.[3] And ABET, the accreditation organization for undergraduate engineering programs in the United States, has required "an understanding of professional and ethical responsibility" only since 2000. Interest in and attention to ethical issues in engineering has grown significantly in recent decades. Many undergraduate engineering programs now require a full semester course on engineering ethics; more than a dozen textbooks on the subject are available; at least one major journal on the topic is well established; issues of professional ethics are commonly featured in the publications of engineering professional organizations; and national and international conferences regularly are devoted to, or include, engineering ethics topics.

Textbooks in the field of engineering ethics take a variety of approaches. Case studies are commonly featured, though the types of cases vary. Some textbooks emphasize prominent real-world cases, such as the *Challenger* disaster or, more recently, the Fukushima Daiichi Nuclear Power Plant accident. Others present short scenarios, sometimes hypothetical, often based on the experiences that practicing engineers report. These cases typically depict individual engineers who are caught in complicated circumstances and must make a difficult ethical decision. Some textbooks emphasize the use of ethical theory or a methodology for solving ethical dilemmas. Although all the textbooks seem to have an overlapping core, topics vary from textbook to textbook. For example, some include intellectual property or research ethics or the rights of engineers, whereas others do not.

This book takes a new approach to engineering ethics. It presents the field as a series of debates on selected topics, both enduring and contemporary. The enduring topics are settled, in the sense that a broad consensus exists on one side of the debate. The contemporary topics are unsettled, in the sense that opinions on both sides are still evolving.

This approach has much to offer. For one, the field of engineering ethics has been largely codified in professional codes of ethics and in established norms of practice. The subject matter can thus be treated as if engineers simply needed to learn a list of rules and follow them. Were this indeed the case, the only challenge that engineering ethics would pose would be how to promulgate the rules.

The reality of engineering practice is that many challenging situations occur for which the right thing to do cannot be specified with a simple answer or known in advance. Often the right thing to do depends on the fine-grained details of the situation, and often the best course of action requires taking one step, seeing what happens, and, depending on what happens, taking this or that next step. Each step is dependent on the response to the prior step. General principles and rules are helpful in such circumstances but by no means sufficient. The debate format is useful for developing the skills and perspectives needed to handle uncertain situations.

The debate format uncovers alternative perspectives, generates reflection, and exercises critical thinking. Readers are called on to consider the reasons for prevailing norms of practice, challenge them, and develop nuanced positions. Importantly, this prepares engineers for a future in which the profession and its norms may evolve.

Of course, I can hear the skeptic saying, "But many of the issues in engineering ethics are not debatable, and it is dangerous to give the impression that they are." "Engineers," the skeptic might argue, "should adhere to their professional codes of ethics and should behave according to the norms of the profession—no questions asked."

Although there is some truth to this argument (for example, engineers should never take bribes, and they should never approve substandard materials), there is also a simple counter. In fact, although general rules and behavioral norms can be clearly stated, these broad prescriptions are often difficult to interpret and apply to complex situations. For example, even though engineers should never take bribes, in some situations bribery may not be obvious. If one exchanges gifts with business associates during the holidays, is that bribery? Or if foreign business associates give engineers trinkets symbolizing their pride in their country, are the engineers accepting bribes? General rules and norms must be interpreted for real-world situations, and in order to do this, engineers must understand the underlying rationale for the rules and norms.

The debate format, finally, is supported by one of the profound insights of the philosopher John Stuart Mill. Mill argued that any idea, even if widely accepted as true, if not actively debated, runs the risk of becoming "dead dogma." That is, if a generally accepted idea is never challenged and discussed, the idea will not be fully understood and will be easily discarded in the face of dissent. As Mill wrote: "However unwillingly a person who has a strong opinion may admit the possibility that his opinion may be false, he ought to be moved by the consideration that however true it may be, if it is not fully, frequently, and fearlessly discussed, it will be held as a dead dogma, not a living truth."[4]

Mill's idea is especially apt for debating issues that are considered settled in the profession. Several chapters of this book present debates that may appear somewhat contrived in that a consensus exists on which side of the issue engineers should take. For example, in asking in chapter 1, "Can ethics be taught?" and in chapter 2, "Does engineering need a code of ethics?" readers can predict that the answer will be yes. However, even though the conclusion of these two chapters is no surprise, it would be a mistake to suppose that the conclusion should simply be proclaimed and accepted without discussion or defense. Understanding why these answers are so widely accepted makes all the difference in the world. For example, unless the question "Does engineering need a code of ethics?" is vigorously debated, the importance of codes of ethics will be taken as "dead dogma, not a living truth."

Mill's argument is less relevant when it comes to less settled contemporary topics. Are whistleblowing engineers heroes or traitors? Will autonomous cars ever be safe enough? Is social justice in the scope of engineers' social responsibilities? There is no agreement on the norms regarding these issues. In the chapters on these issues, the debate format provides a useful framework for exploring and reflecting. These chapters are designed to provide concepts and arguments that help readers sort through the issues, considering multiple factors and forming enlightened positions on matters that cannot be resolved easily or may not at this time be resolvable.

The eight chapters of this book do not treat every possible topic in engineering ethics. In selecting the subjects to cover, I had several considerations in mind. I wanted to include issues that are considered the core of the field and would be included in any textbook on engineering ethics, such as the nature of the profession, codes of ethics, the nature of ethical reasoning, and whistle-blowing. I also wanted to include both micro- and macro-ethical issues. Micro-ethical issues are those that arise for individual engineers in their daily lives. Here the issues often take the form of asking what an engineer caught in a difficult situation ought to do, considering the engineer's personal and professional responsibilities. In contrast, macro-ethical issues are those that involve the engineering profession as a whole or are social and policy issues involving technology and/or the work that engineers do. For example, deciding when autonomous cars will be safe enough (the topic of chapter 7) and whether to hold engineers responsible for social justice (chapter 8) are issues that must be dealt with on a broad scale and cannot simply be addressed through the behavior of individual engineers. There must be collective agreement and action or social policies to handle these issues.

After the topics were selected, the structure of the book fell into place. The first three chapters comprising Part I can be thought of as foundational issues. Can engineering ethics be taught? Do engineers need a code of ethics? How should engineers think about ethics? Each chapter frames an enduring debate, and together they set the scene for what follows. Part II consists of two chapters, each focusing on a relationship that practicing engineers have in their day-to-day practice. Most engineers are employed in organizations—small or large companies or government agencies—in which they are accountable directly or indirectly to their employer. Either as employees or independently as private consultants with their own businesses, engineers also work with clients. Chapters 4 and 5 are devoted respectively to ethical issues arising in engineer-client relationships and engineer-employer relationships. These chapters deal primarily with micro-ethical issues. Part III takes up macro-ethical issues. Returning to an enduring question, chapter 6 asks whether it is bad individuals or bad organizations that lead to technological mishaps. This chapter uses real cases of engineering accidents and wrongdoing in which the question of responsibility points to bad individuals and/or bad organizational cultures. Chapter 7 is focused on safety, one of the most important topics in engineering ethics. However, in chapter 7, safety is taken up in the context of an emerging technology, autonomous cars. The book concludes with a topic that may be even more controversial than the others. To ask whether there is a connection between engineering and social justice is to ask an extremely complicated question requiring a deep understanding of the relations among technology, society, and engineering. It is also to suggest that engineers may be responsible for more than is traditionally thought to be a part of engineering. Chapter 8 includes a discussion of environmental justice and attempts to bridge macro- and micro-level analysis.

Chapter 1 needs special explanation. At first glance, some may think that this chapter, which asks whether engineering ethics can be taught, is more appropriate for instructors of engineering ethics than for engineers (engineering students or practicing engineers) because it discusses the goals of engineering ethics education. I have included this chapter and placed it first with another purpose in mind. My experience teaching and interacting with engineers is that some are, at first, quite skeptical about the field of engineering ethics. Starting the book with a chapter asking whether engineering ethics can be taught is a way of confronting and addressing that skepticism right away, so that it doesn't get in the way of subsequent chapters. The chapter also serves another purpose—an important purpose for those who are not skeptical. Answering the question whether engineering ethics can be taught leads directly to a discussion of the goals of teaching ethics, and this provides an opportunity to reflect on what can and cannot be accomplished in a course or a book on the topic. Importantly, that discussion sets realistic expectations for the chapters and ideas that are to follow.

PART ONE

Foundational Issues

1 CAN ENGINEERING ETHICS BE TAUGHT?

IN 2010, after a two-year inquiry, a judge concluded that Canadian prime minister Brian Mulroney had acted inappropriately when he accepted large amounts of cash from a German Canadian arms lobbyist. The judge suggested that all public servants should get ethics training. Peter Worthington, a columnist for the *Toronto Sun*, responded to this suggestion in the following way: "A case can be made that 'ethics' are something that you either have, or you don't have. Or, to put it slightly differently, ethics are a code you subscribe to or chose to ignore for reasons of personal interest. . . . All the training, teaching, studying, reading, or lectures in 'ethics' will not make a person more ethical if he or she does not have these core values to begin with."[1]

Reading between the lines, we might think that Worthington believes people acquire their core moral values during their childhood. Once they reach a certain age, not much can be done. If a person was brought up to have ethics, great; if, on the other hand, someone didn't have that kind of upbringing, forget it—the person will never change and never learn to become ethical.

Worthington expresses a form of skepticism that is not uncommon when it comes to teaching ethics to undergraduate engineering students. For example, Karl Stephan, a professor of engineering, described the following encounter: "Some years

ago I argued with a fellow professor about the issue of engineering ethics education at the college level. His point was along
the lines of, 'Hell, if eighteen-year-old kids don't know right from
wrong by the time we get 'em, they're not going to learn it
from us.' "[2]

Despite such skepticism, many (if not most) undergraduate
engineering programs in the United States, as well as in many
other countries, require training in ethics as part of the curricu-
lum. One reason they do so, is that ABET, the accreditation organization for undergraduate engineering programs in the United
States, requires it.[3] ABET specifies a list of outcomes that students
must achieve by the time they graduate. These include "an understanding of professional and ethical responsibility" and acquiring
"the broad education necessary to understand the impact of engineering solutions in a global, economic, environmental, and
societal context." Programs demonstrate their achievement of the
outcome in a variety of ways, including semester-long courses on
engineering ethics.

If the skeptics are right, ABET's requirements about ethics
are a waste of time. If ethics cannot be taught, there is no point
in requiring that engineering programs teach it. So, the question
whether ethics can be taught is important for engineering as well
as more broadly.

The question is not a simple one, at least not as simple as the
skeptics suggest, and in exploring it in this chapter, we will necessarily go deeper into the goals of ethics education. We can, for a
start, dismiss the idea that the goal of ethics education—in engineering as well as other fields—is to ensure that no one—not one
single engineer—will ever do anything wrong. Skepticism is appropriate for this impossible goal. Yet there are more modest goals
that, if achieved, would increase the likelihood of individuals
behaving ethically. For example, in engineering one goal might
be to inform engineering students about the codes of ethics promulgated by engineering professional societies. Ensuring that
students know about these codes and that they are considered
important by engineering professional organizations increases
the likelihood that students will follow them.

Some readers may be surprised that the first debate in a book on engineering ethics is whether ethics can be taught. Indeed, some may think that this chapter is only for ethics instructors. However, the chapter is placed first for a purpose: it sets expectations for what is to follow. Sorting out what can and can't be taught and what are and are not appropriate goals for ethics education prepares readers for what will follow in subsequent chapters.

MOTIVATIONS FOR MORAL BEHAVIOR

Although this chapter focuses on the question whether engineering ethics can be taught, we begin with a debate about what motivates moral behavior. At the core of the question about whether ethics can be taught is a set of beliefs about the sources or causes of moral behavior. In other words, in order to answer the question about what can be taught, we must ask a prior question about what goes on in a person when he or she makes a moral decision. Skeptics seem to eschew this question by insisting that what motivates moral behavior is something in individuals that is fixed and not amenable to influence—either ever or after a person has reached a certain age. By contrast, ethics education presumes that ethical behavior can be influenced, and it presumes that the target—the thing to be influenced—is how people make moral decisions. Ethics education can make people better, the presumption is, by improving how they make moral decisions. Many believe that reasoning is a significant part of this; hence, ethics courses are often designed to improve ethical reasoning. Curricula and courses aim to provide students with information, ideas, and experiences that will improve their ethical reasoning skills. In fact, many ethics courses focus on ethical theories such as utilitarianism, deontology, and virtue ethics precisely because these theories are thought to be tools and heuristics for improved ethical reasoning. (These theories will be discussed in chapter 3.)

Recently, however, the focus on ethical reasoning has been challenged. Critics claim that ethical behavior is based largely on moral intuitions, not reasoning. Moral intuitionists (also referred to as social intuitionists because they believe that moral intuitions are interpersonal) claim that if we look at how people make moral decisions, we will find that intuitions play a much larger role than previously thought. Their position is aligned with skeptics insofar as it suggests that teaching moral reasoning will have little or no effect. Moral intuitionists argue that reasoning comes in only after intuitions; reasoning is, they say, more like rationalization in the sense that it is used to explain and justify what one's intuitions tell one to do. Jonathan Haidt put it succinctly when he wrote, "The central claim of the social intuitionist model is that moral judgment is caused by quick moral intuitions and is followed (when needed) by slow, ex post facto moral reasoning."[4]

Moral intuitionists often point to human experience as evidence for their position; they point to individuals who are repulsed by the sight of certain kinds of behavior. For example, they claim that without reasoning individuals find the idea of incest or torture of children abhorrent, and that for some the thought of eating animals is viscerally distasteful. These experiences suggest that a person's ethics is based on her or his intuition, not reasoning. This view is in line with skepticism about teaching ethics in the sense that it claims that ethical behavior is the result of something somewhat primitive in humans, something not amenable to reason.

Nevertheless, there are some serious criticisms of moral intuitionism, criticisms that take us deeper into the matter of what motivates moral behavior. One counter is that moral intuitions don't come out of nowhere; they aren't inborn. Reasoning comes into play as individuals form their moral intuitions. According to this line of thinking, individuals acquire their intuitions through developmental processes that involve reason. In other words, one's experiences and reasoning about those experiences shapes one's intuitions. If this is right, then addressing moral reasoning in education might lead to better moral intuitions.

Another line of criticism of moral intuitionism points to some obvious cases in which reasoning seems to change a person's moral intuitions. For example, through discussion and debate with a friend, one might be convinced to become a vegetarian, or one might, through reflection, decide to refrain from telling certain kinds of jokes, say, because they perpetuate racial or gender stereotypes. The point here is that whatever role intuition may play in ethical decision-making, it can be tempered by reasoning.

Yet another criticism points to situations that are not so simple (and arguably more common) in which the right thing to do is more complex and challenging. As David Pizzaro and Paul Bloom explain, Haidt "was likely correct that we do have quick and automatic responses to certain situations—killing babies, sex with chickens, and so on—and although these responses can be modified and overridden by conscious deliberation, they need not be. But most moral cognition is not about such simple cases; in the real world, moral reasoning is essential."[5]

In these more complicated situations, intuitions may not help. Imagine having to decide whether to blow the whistle on your boss for apparently illegal behavior or deciding whether to do what a friend has asked even though you think the friend is, at the moment, not mentally stable. In such situations, individuals are compelled to think carefully about the details and the consequences of taking this or that line of action. In these circumstances, being able to think through the situation, reason about alternative options, and anticipate consequences is not just helpful but essential.

Although Pizarro and Bloom's point seems correct, as do the other criticisms of moral intuitionism, moral intuitionists seem right to insist that moral behavior involves more than reason alone. Unfortunately, the debate cannot be resolved here. Indeed, the dialogue between moral intuitionists and moral reasoning advocates is a recent manifestation of a philosophical dialogue that has been going on for several thousand years. In earlier manifestations, the question was whether reason is a slave to the passions or vice versa. The point of exploring this debate is to show that the question whether and how ethical behavior can be influenced is a deep and enduring matter and not so easily dismissed as the skeptics would have it.

We turn now to another way of answering the question. We can look at what can and is being done to teach ethics. After considering what ethicists take to be the goals of teaching engineering ethics, we will be in a better position to determine whether education can influence moral behavior.

CAN ENGINEERING ETHICS BE TAUGHT?

We can begin by considering the idea that engineering ethics may have a knowledge component. That is, setting aside both reasoning and intuition, consider that being an ethical engineer may require information. If so, then there is at least one aspect that can be transmitted by means of education—that is, this one component can be taught.

We know that engineering requires mastery of a body of knowledge. That is, to do much of the work that engineers do, one needs technical knowledge and know-how. The knowledge or know-how needed depends on the nature of the project and the field of engineering, but engineering—at least modern engineering—is an activity that requires a good deal of expert knowledge. Although this is not often conceived as part of engineering ethics, in fact competence in engineering is essential to ethical engineering. For example, one can't build a safe bridge, identify environmental pollution, or design life-saving biomedical devices unless one has the technical expertise. In fact, many codes of ethics of engineering professional organizations specify that engineers should maintain and improve their technical expertise and should only perform jobs in areas of their competence. This is important to note: competence in engineering is part of ethical engineering, so technical knowledge is part of engineering ethics.

Nevertheless, technical knowledge and expertise do not alone or necessarily lead to ethical engineering. Engineering ethics is directed at how technical competence is deployed and at understanding the contexts and relationships in which it is used. This requires knowledge, but of a different sort. For a start, the knowledge component of engineering ethics involves knowledge of professional norms and practices; knowledge of the expectations of employers, clients, and the public; knowledge of how to manage these relationships; and knowledge of the social consequences of technical endeavors.

To better understand how knowledge is relevant to engineering ethics, we can consider a case from the National Society of Professional Engineers' (NSPE) repository of cases. The NSPE has constituted a Board of Ethical Review (BER) consisting of seven licensed members appointed by the society's president, and each year the board reviews cases that have been submitted. Whether the cases are real or hypothetical is not known. The BER analyzes each case and states whether the behavior described in the case is ethical or unethical. The analysis is meant in part to interpret the NSPE Code of Ethics. (Engineering codes of ethics will be detailed in chapter 2.)

Consider NSPE Case 15-10:

> Engineer A, a professional engineer, works as the director of the local government building department. Engineer A also has a part-time sole engineering practice and prepares a set of structural engineering drawings for Client X. The drawings must be approved by the local building department. Engineer A does not participate in the review or approval of the drawings but Engineer A's assistant, Engineer B, a professional engineer, reviews and approves the engineering drawings prepared by Engineer A.[6]

In reviewing this case, the BER concludes that both Engineer A and Engineer B behaved unethically.

What exactly did the engineers do wrong and why? Engineer A seems vaguely aware of the concept of conflict of interest and entirely unaware of how to handle such situations. A conflict of interest is when a person has a role that requires the exercise of judgment on behalf of others but also has an interest that might taint, or merely appear to taint, his or her judgment. (Conflicts of interest will be discussed in more detail in chapter 4.) In this case, Engineer A is expected to act as a representative of the local government building department—that is, he is expected to act in that department's interests. However, he has a personal interest in having the drawings approved, and that personal interest has the potential to influence his judgment. Hence, he should not be the one who reviews the drawings. The objectivity of his judgment could easily be called into question.

Engineer A seemed to understand that there was a problem in his reviewing his own plans because he asked his assistant to do the review. However, in doing this, Engineer A made things worse in two ways. First, he didn't inform his employer about the conflict of interest. The BER describes Engineer A's behavior as a failure to "fulfill his obligation to act as a faithful agent or trustee to his employer." Second, Engineer A compounded the problem by asking someone he supervised to review the drawings. The BER is unambiguous in specifying that Engineer A should not have had Engineer B review the drawings because Engineer A supervised Engineer B. The BER explains, "It would be unethical for Engineer A to provide the services in the manner indicated, even if he had obtained approval from his supervisor, because he cannot require his subordinates to approve his work." The BER insists that a subordinate is not capable of rendering objective (untainted) judgment on a supervisor's work.

This case suggests the importance of knowledge to engineering ethics in that Engineer A might have behaved differently had he known about the NSPE Code of Ethics, what constituted a conflict of interest, and what the norms of professional practice are for someone in his position. Of course, it is possible that Engineer A knew all of this and still did what he did. In this respect, knowledge does not make a difference to someone who is willfully unethical. Nevertheless, for those who wish to do the right thing, knowledge is crucial. Individuals are less likely to behave unethically if they are aware of their ethical obligations—that is, the norms and practices in their field.

The same could be said for Engineer B. She might have behaved differently had she known about norms of professional practice. Again, we can't be sure whether Engineer B was merely ignorant of professional norms or knew the norms but felt pressured to violate them because of her supervisor's request. If it was ignorance, then knowledge of professional norms and training in how to deal with difficult situations might have helped her respond to Engineer A's request. If, on the other hand, she knew she shouldn't do the review but felt pressured, she could have pointed to the NSPE Code of Ethics and would have been better equipped to articulate why she had to refuse Engineer A's request.

This case illustrates that knowledge (not just technical knowledge) has a role to play in engineering ethics. It can make a difference in how people behave. In this case, knowledge of engineering professional codes and norms of professional practice might have affected the behavior of both engineers. In other cases, other kinds of knowledge might be relevant to an engineer's decision. To be sure, as already indicated, perhaps both engineers had the knowledge and chose to ignore it. For them, something more than knowledge would be needed to discourage their unethical behavior.

Next we'll delve more deeply into the knowledge component of engineering ethics and see that engineering ethics education can do more than provide knowledge.

THE GOALS OF TEACHING ENGINEERING ETHICS

Within the burgeoning literature on engineering ethics education, as well as professional ethics education more broadly, there is much debate about what can and should be taught and how best to teach it. For example, some think that the primary goal should be to make students aware of the ethical issues that might arise in professional practice. Others think that the goal should be to improve ethical decision-making and judgment. Yet others think that the emphasis should be on motivating and even inspiring students to behave ethically. Of course, these goals are not mutually exclusive. Four types of goals come up persistently in the literature. Engineering ethics education should: (1) provide knowledge of codes and standards; (2) increase awareness of ethical issues and develop the ability to identify ethical issues; (3) provide training in ethical decision-making; and (4) inspire the motivation to be ethical.

Of course, skeptics might still insist that none of these goals is achievable because education can't counter poor moral upbringing or bad character. For that reason, it is important to examine each goal more fully and consider the effect that each aims to achieve.

Knowledge of Codes of Professional Conduct

The discussion of NSPE Case 15-10, above, illustrates the first goal. This case shows that knowledge of a professional code of ethics and, more broadly, norms of professional practice might make a difference in how engineers behave. However, more than an awareness of codes of ethics and standards of behavior may be needed. The codes of ethics and standards of behavior must be understood. They must be grasped in a way that allows them to be used in or applied to particular situations. In other words, although familiarity with codes of ethics is essential in preparing for the ethical challenges that engineers face as professionals, exposure and familiarity are not enough. Engineering ethics education should provide training in how to interpret and apply the statements in the codes.

To better understand the importance of learning to interpret and apply standards and codes of ethics, consider a criticism put forward by Stefan Eriksson, Gert Helgesson, and Anna Höglund.[7] These authors focus on health care ethics, but their critique applies to all professional ethics. Their concern is with an overreliance on codes of ethics combined with insufficient attention to the problem of interpreting them. Simply knowing what a code of ethics says doesn't help people in real-world situations, because in order to see how a code or rule applies to a situation, the code or rule must be interpreted. For example, in Case 15-10, Engineer A may have known that the NSPE Code of Ethics specifies that engineers are to act as faithful agents of their employers, but he may not have made the connection between having his drawings approved and being faithful to his employer. He may not have thought about how his employer's interest might be in tension with the interests of his private practice clients. Simply put, he might not have really understood the edict to be faithful to one's employer.

Consider a different situation that can arise in engineering practice. Many engineering codes of ethics specify that engineers should act faithfully on behalf of their clients. Imagine an engineer who discovers a safety problem while working on a client's building. The safety problem could affect people who work in the building, say, by increasing the risk of fire. The engineer tells her client and the client asks the engineer to keep quiet about the safety issue because the client is planning to sell the building. Although the codes of ethics make it clear that engineers have a responsibility to their clients, the codes also specify that engineers should hold paramount the safety, health, and welfare of the public. In this situation, simply knowing that a code of ethics exists and knowing what it says is not enough to help the engineer figure out what to do. She may well know of the commitment to hold the safety of the public paramount as well as the commitment to be faithful to clients and still not know how to handle a situation when two professional norms seem to conflict.

Of course, there may be no simple answer to the engineer's dilemma, but familiarity with similar situations and practice thinking through those situations might help the engineer. Charles Abaté argues that students should be required to engage in activities that involve interpreting and debating the meaning of the codes so that they will have experience in how to apply them.[8] Moreover, in practicing how to interpret and apply particular norms, students learn strategies that can be used in similar situations. Engineering education should, therefore, give students opportunities to grapple with the meaning and implications of statements in codes of ethics and practice interpreting and applying them.

Another problem with knowing only that codes exist and not what they mean and how they apply is that the lack of understanding may lead to legalistic thinking. This is a problem both because law and ethics are not the same and because legalistic thinking may suggest that one should simply follow the rules without consideration of aspects of a situation that aren't covered. Codes of ethics are intended to provide guidelines and not to be rules that one must follow no matter what.

The criticism of teaching codes of ethics made by Eriksson, Helgesson, and Höglund is not, then, a rejection of the goal of raising awareness of codes of ethics; rather, it is an argument for doing this in a way that develops skill at interpreting and applying the codes and doesn't lead to an overemphasis on legalistic thinking.

Awareness and Ability to Identify Ethical Issues

Engineering codes of ethics are critical to engineering and to engineering ethics, but not all ethical issues can be covered in a a code. Another important goal of engineering ethics education is to cultivate the ability to identify ethical issues. Consider the following case.[9]

Joe is employed as a manufacturing engineer. In this role, he meets periodically with vendors who supply his company with materials. It is common to chat informally with vendor representatives, and Joe discovers that one of the representatives—Frank—is an avid golfer, as is Joe. During one of their meetings, Frank invites Joe to play golf at a posh country club that Joe knows has an excellent golf course. Joe takes Frank up on the offer, and they have a competitive but friendly match with a few other players as well. Frank invites him again and suggests that they play for small amounts of money. Although Joe is reluctant, he agrees. The practice of playing golf together continues; Joe is surprised, but he often wins money on the games. After some time, Frank nominates Joe for membership in the country club, and his membership application is approved.

This continues for several years, and then Joe is called into a meeting and told that his company is having financial difficulty and needs to institute cost-cutting measures, including termination of several vendor contracts. Joe is assigned to work with two other engineers to decide which vendors will be cut. Because of his relationship with Frank, Joe is now worried about his objectivity in making this decision. He tells the other two engineers about his relationship with Frank and continues to work with them. Joe concludes that it would be best to eliminate Frank's company; the other engineers independently arrive at the same decision.

Joe decides that he must tell Frank what has happened, so he meets with Frank and tells him the news. Frank is quite upset and wants to know what Joe did to try to protect Frank's company. Frank becomes angry and blurts out: "I don't believe this! What kind of friend are you, anyway? Didn't I get you into the country club? And how good a golfer do you think you are, anyway? How do you think you've won all that money from me over the years? You don't really think you're that much better at golf than I am, do you?"

In this case, Joe was unaware of the pitfalls of establishing a personal relationship with a vendor. He didn't recognize that entering into a relationship outside work might take him into delicate ethical territory. He didn't think to question Frank's motives for increasingly involving him in a golfing relationship, nor did it occur to him that he might someday have to make a tough professional decision that would affect Frank. This is a case in which the engineer might have benefited from having explored the ethical issues that can arise in professional relationships. Joe might have picked up on Frank's motives sooner or anticipated potential problems had he been more aware of the nuances of professional relationships. The case is intriguing because it is difficult to determine when Joe went too far, if ever. Should he have recognized a potential problem sooner? Should he have said no to the invitation to play golf? To play for money? To join the golf club?

Analysis of case studies like this can raise awareness of ethical issues and sharpen one's ability to identify ethical issues before they become difficult and complicated. Exposure to case studies—whether real or fictitious—increases the likelihood that a new engineer will identity ethical issues quickly and early on before they cause too much trouble. Abaté argues that analysis of case studies is effective because it draws on a basic human capacity for pattern recognition.[10] Exposure to case studies allows one to discern patterns that can then be seen in other, real-world situations.

Another strategy that helps to identify ethical issues is familiarity with moral concepts and frameworks. The idea here is that when one has a vocabulary for thinking about ethics, one is more likely to recognize situations involving ethics. Thinking carefully about moral concepts such as integrity, loyalty, respect, or conflict of interest and their implications increases the likelihood that one will see the relevance of these concepts to real-world situations. For example, in chapter 3, when we discuss ethical concepts and theories in more detail, we will explore the concept of never treating a person merely as a means. One is much more likely to notice when someone is being treated in that way when one has discussed this concept and has had experience in using it. Ethical concepts and theories provide language, principles, and ways of looking at the world that help people notice what they might otherwise ignore.

Moral Reasoning and Improved Decision-Making

The ability to identify ethical issues is an important part of being an ethical engineer. However, it is one thing to see that there is an issue and quite another to know what to do about it. A central focus of engineering ethics is on reasoning and ethical decision-making. This takes us back to the skeptics, moral intuitionists, and questions about what is going on inside someone when he or she makes a moral decision.

Without a doubt, engineers often face situations in which they must make tough moral decisions: Should I blow the whistle on my employer? Should I violate confidentiality when my client appears to be engaged in illegality? What should I do if asked to pay to play in order to get a contract? Should I let this design go to the next level of development when I believe that the design will never be safe enough to be built? So, the question isn't whether engineers engage in moral decision-making. They do! The question is whether and how ethical decision-making can be taught.

Since engineering ethics education is relatively new, there is still some uncertainty about how best to teach it, and in this context, ethical decision-making poses the greatest challenge.[11] The standard approach is to provide practice—practice with feedback and discussion. Here again case studies and case analysis are considered the best approach. Courses that provide the opportunity to consider, reflect upon, analyze, discuss, and even debate what to do in tough situations are thought to improve decision-making skills. Engineering ethics courses often include case or scenario analysis and role-playing, providing students an opportunity to experience ethical dilemmas and reflect on how to handle these situations. The idea is that calm and careful reflection on situations and hearing the ideas of others will prepare engineers for the real world in which there may be pressure to act quickly and fewer opportunities for discussion and reflection.

Motivation

Another goal of engineering ethics is to inspire and motivate ethical behavior. The argument for this component is straightforward. Providing people with knowledge, skills, and experience isn't enough to ensure that they will act on the knowledge or exercise the skills. They must be motivated to be willing to use their knowledge and skills especially in challenging situations, circumstances that may require acting against their self-interest.

Little is formally known about how best to inspire and motivate. Yes, we all have had the experience of being inspired by a charismatic leader, a passionate speech, or a life-changing event, but how can this be translated into engineering education? Some ethicists have suggested that it might be done by exposure to exemplary and heroic engineers.[12] The idea here is that hearing about engineers who have made great sacrifices to do the right thing or who have taken on noble causes or who have withstood pressure to engage in wrongdoing will inspire other engineers to do likewise. William LeMessurier is often put forward as such an inspiring figure. LeMessurier was a distinguished structural engineer and a consultant in the development of the innovative Citicorp Headquarters building in New York. After the building was completed (in 1977), LeMessurier learned that it did not meet the safety standards for buildings situated as it was. As Michael Pritchard explains, LeMessurier "knew how to correct the problem, but only at the cost of a million dollars and at the risk of his career if he were to tell others about the problem. He promptly notified lawyers, insurers, the chief architect of the building, and Citicorp executives. Corrections were made, all parties were cooperative, and LeMessurier's career was not adversely affected."[13]

LeMessurier is inspiring because he took on personal risk in order to ensure the safety of a building that he had designed. Several engineering ethicists have compiled stories of heroic and exemplary engineers who are less well known than LeMessurier but equally impressive in how they risked their own well-being in order to protect others.[14]

Inspiration is related to moral courage. Engineers often find themselves in situations in which their work environment makes it difficult for them to express their ethical concerns: a concern about safety, an awareness that vulnerable populations will be harmed by a project, perhaps a suspicion that something illegal is going on. Moral courage is the ability to act for moral reasons and despite one's fear or the likelihood of personal harm. Moral courage is the capacity to speak out and to stand up for what one thinks is the right thing to do. It may well be that most people want to do the right thing but not all people have the moral courage to act in tough situations. Inspiration may well promote moral courage.

CONCLUSION

Can engineering ethics be taught? The skeptics who answer this question in the negative seem to have an overly simplistic picture of human behavior. To suppose that ethics can only be taught in childhood and that nothing after that can affect a person's moral thinking or moral behavior seems not to acknowledge or appreciate how people engage in ethical decisions. Moreover, it seems misguided to suppose that ethics is simply a matter of learning a few rules in early childhood and then following them. To be sure, learning a few basic moral principles early on in life is probably a good thing, but real-world situations are often complex and require more than following general rules. Rules need to be interpreted, and it is not always easy to see which rule applies when and how.

In this chapter we have seen that engineering ethics has an informational or knowledge component (for example, know what the norms for engineering are); a cognitive component (for example, recognize when there is an ethical issue); a reasoning component (for example, make a moral judgment); and a motivational component (for example, move yourself to act). All four dimensions can be addressed through education.

Engineering ethics can be taught by providing knowledge of professional codes and how to interpret and apply these codes to real-world situations. Awareness of ethical issues can be heightened, as can the ability to identify ethical issues. Training and experience in ethical decision-making can be provided. And engineers can be inspired to want to be ethical and to act in morally courageous ways. Although all of this may not make so-called bad people good or ensure that all engineers will always do the right thing, pursuit of these activities increases the likelihood that engineers will be better able to handle the ethical issues that arise in their professional lives.

This chapter has focused on individuals, specifically, on whether education can influence the behavior of individuals. However, engineering ethics is not just a matter of individual behavior. It is also a matter of institutions, organizations, and environments enabling individuals to behave in certain ways. Organizations and institutions can do a lot to promote and facilitate ethical engineering. In the next chapter we turn our attention to engineering as a profession and consider the activities of engineering professional organizations and codes of ethics in promoting and facilitating ethical engineering.

SUGGESTED FURTHER READING

Articles in *Science and Engineering Ethics*

Articles in *Journal of Professional Issues in Engineering Education and Practice*

Barry, Brock E., and Joseph R. Herkert. "Engineering Ethics." Pages 672–92 in *Cambridge Handbook of Engineering Education Research*. Cambridge: Cambridge University Press, 2015. https://doi.org/10.1017/CBO9781139013451.041.

National Society of Professional Engineers, BER Cases, http://www.nspe.org/resources/ethics/ethics-resources/board-of-ethical-review-cases

2 DO ENGINEERS NEED CODES OF ETHICS?

M OST engineering professional organizations, in the United States and globally, have adopted codes of ethics and professional conduct. For example, the National Society of Professional Engineers, the American Society of Civil Engineers (ASCE), the British Royal Academy of Engineering, and the World Federation of Engineering Organizations all have codes of ethics. Each code has a history and reflects a process wherein engineers gathered, reflected on their fundamental values as engineers, and articulated the principles by which members should conduct themselves. Many of the older codes have changed over the years, reflecting changes in the group's view of its role and responsibilities. Although the existence of codes and their content are currently not widely controversial, skeptics can and have challenged the need for and value of codes of ethics. Some argue that codes of ethics are not necessary; others argue that they are counterproductive insofar as they lead to minimal standards of behavior; yet others argue that they are ineffective because they have no enforcement power. In this chapter, we will explore the value of engineering codes of ethics by framing a debate about whether codes of ethics are necessary. The debate requires us to understand the nature of engineering as a profession. Engineering codes of ethics play an important role in the engineering profession, and they reflect the complex role of engineering in society.

Each field of engineering has its own code of ethics, so there is no single code of ethics that applies to all engineers. Nevertheless, the fundamental principles expressed in each of the codes tend to be similar in content. The exception to this is concerns that are unique to a domain of engineering. For example, sustainability is emphasized in civil and environmental engineering codes of ethics, and patient rights are a focus in biomedical engineering codes. The similarity of codes in the fundamental principles they express is partly because principles must be broad in scope to apply to the wide range of situations in which individual engineers can find themselves. At this broad level, the principles are somewhat generic.

The Fundamental Canons of the NSPE Code of Ethics illustrate this generality and show how the edicts in a code can be broadly applicable. The NSPE Fundamental Canons are as follows:

> Engineers, in the fulfillment of their professional duties, shall:
> 1. Hold paramount the safety, health, and welfare of the public.
> 2. Perform services only in areas of their competence.
> 3. Issue public statements only in an objective and truthful manner.
> 4. Act for each employer or client as faithful agents or trustees.
> 5. Avoid deceptive acts.
> 6. Conduct themselves honorably, responsibly, ethically, and lawfully so as to enhance the honor, reputation, and usefulness of the profession.[1]

After the Fundamental Canons, a list of Rules of Practice is provided. The rules spell out what each of the canons means more specifically though even the Rules of Practice are fairly general. They in turn are followed by a list of Professional Obligations, which spell out in further detail the meaning of the Fundamental Canons. Again, even here the statements are general in scope. For example, the first of the Professional Obligations is: "Engineers shall be guided in all their relations by the highest standards of honesty and integrity."

Statements very similar to the NSPE's Fundamental Canons are found in many other codes of ethics. For example, the ASCE Code of Ethics contains an edict that is almost identical to the first canon of the NSPE code: "Engineers shall hold paramount the safety, health and welfare of the public and shall strive to comply with the principles of sustainable development in the performance of their professional duties."[2]

The code of the IEEE (Institute of Electrical and Electronics Engineers) contains something similar:

> We, the members of the IEEE, in recognition of the importance of our technologies in affecting the quality of life throughout the world, and in accepting a personal obligation to our profession, its members, and the communities we serve, do hereby commit ourselves to the highest ethical and professional conduct and agree:
>   1. to hold paramount the safety, health, and welfare of the public, to strive to comply with ethical design and sustainable development practices, and to disclose promptly factors that might endanger the public or the environment.[3]

And the American Society of Mechanical Engineers' (ASME) code also contains a comparable statement:

> Engineers uphold and advance the integrity, honor and dignity of the engineering profession by:
>   1. Using their knowledge and skill for the enhancement of human welfare.[4]

In addition to being similar regarding protecting public safety and welfare, engineering codes of ethics tend to be alike in specifying that engineers have obligations to clients and employers. Engineering practice involves interactions with a variety of constituents, including clients and employers but also contractors, regulatory agencies, investors, and other engineers. Codes vary in how they address these other constituents, whether implicitly or explicitly.

The principles expressed in engineering codes of ethics make it clear that becoming an engineer involves more than simply pursuing one's self-interest. It involves taking on a variety of commitments that go beyond self-interest and may well require self-sacrifice in extreme situations. For example, statements in some codes entreat engineers to speak out when they are aware of threats to public safety even when it will anger or lead to retaliation from a client or employer.

So, what could possibly make codes of ethics controversial or debatable? Well, one criticism is that codes of ethics are generic platitudes that provide no meaningful guidance on how to act. They are too general to be of use to engineers in real-world situations, and sometimes statements in the codes appear to conflict with each other. Those who take this view usually say that codes of ethics are at best designed merely to promote the public image of the profession. They are like "Mom and apple pie," telling the world that engineers will do all the ideal things one would want them to do.

Another criticism, mentioned earlier, is that codes are ineffective because they lack enforcement power. Codes of ethics are promulgated by professional organizations, and rarely, if ever, is an engineer expelled from a professional organization for failure to live up to a code. Moreover, even if an engineer were expelled, it is unclear whether this would affect the engineer's career in any significant way since membership in a professional organization is not a requirement for most jobs.

Note, however, that this criticism does not apply to codes of ethics associated with state licensing boards. As will be explained later, in certain fields of engineering, a license is required to provide certain services. Licensing is done through state licensing boards, and they generally require commitment to a code of ethics. These codes are enforced through disciplinary action by the state board. An engineer's license can be revoked or suspended for a length of time, and engineers can be required to pay penalties. Arguably, this division between the codes of ethics associated with licensing and those promulgated by professional organizations suggests a weakness in the organizations.

Another criticism of professional codes of ethics harks back to chapter 1. If you believe that ethics can't be taught or if you believe that an individual's character is developed early in life and can't be changed after that, then you will likely also believe that codes of ethics will have no effect on those engineers who never had or never developed an ethical character. This argument is related to the lack of enforcement power. That is, whereas a threat of punishment might deter someone who has no internal sense of ethics, without enforcement power, codes of ethics won't influence those individuals.

Yet another argument is that codes of ethics are unnecessary because most engineers work in organizations (companies, government agencies) that have codes of ethics. Engineers, it can be argued, need only follow the code of ethics of the company or agency for which they work. A code of ethics for engineers complicates engineering practice by requiring that engineers adhere to two different codes.

In order to evaluate these criticisms, codes of professional ethics must be put in context and understood as part of the profession of engineering. That engineering is a profession is important because codes of ethics are an important part of the process of professionalization and maintenance of professionalism.

THE PROFESSION OF ENGINEERING

When engineering students complete their undergraduate education, they become not just college graduates but engineers. Obtaining an undergraduate engineering degree is not like getting a degree in physics or history. History majors don't, on graduation, become historians, nor do physics majors become physicists. Yet engineering graduates are engineers as soon as they obtain the undergraduate degree. Indeed, those who graduate with engineering degrees generally think of themselves as engineers even if they work in jobs that don't involve engineering. In this respect, an undergraduate engineering degree is more like a law degree or a degree in medicine. When you complete the requirements, you become a member of the profession. Of course, it is more complicated than this, since lawyers and doctors must complete graduate training, pass a state exam, and take an oath before they can practice, but with some exceptions, that is not the case in engineering. Nevertheless, engineering is one of a small number of occupations that are considered professions.

In the twentieth century, sociologists began studying professions as a distinctive set of occupational groups. They sought to understand why professions emerge and what allows them to acquire and maintain their power and status. Early work on professions emphasized the difference between occupational groups that were organized, fulfilled an important social function, and had some control over their work *and* occupational groups that did not require special education and whose members worked in contexts in which they had little power or control over their work. Members of the latter group were seen as employees, individuals in roles in which they are expected primarily do what their employers tell them to do. By contrast, professions were seen as occupational groups that were organized and had some autonomy in their work.

To some extent the distinction that was made between professions and mere occupations followed a pattern in which those who were seen as professionals worked in private practice (since private practice allowed a high degree of autonomy), whereas those who were employed by others were seen as having an occupation. Those who held jobs working for others, jobs such as carpenter, plumber, auto mechanic, salesperson, or insurance agent, were not seen as professionals. The big differentiating factor here seemed to be autonomy. This included collective autonomy as well as individual autonomy—that is, professions such as medicine and law were seen as having autonomy collectively insofar as they had organizations representing their members, organizations that could make decisions about what the group as a whole would and wouldn't do. Collective autonomy can, in this respect, be seen as a form of self-regulation. The organization can set standards and even control admission to its group. For example, ABET, the organization mentioned in chapter 1 that accredits undergraduate engineering programs, is an organization of engineers. Engineers decide what criteria engineering programs must meet, and engineers review institutions to determine if they have met the criteria. So, ABET is an example of the collective autonomy of engineers.

In the nineteenth century, engineers were seen as professionals or at least as part of an emerging profession because they were primarily engaged in private practice and had a fairly high degree of individual autonomy. Several of the major engineering professional organizations of today—ASCE, ASME, and the American Institute of Mining and Metallurgical Engineers—were established in the nineteenth century. However, at the beginning of the twentieth century, engineers began to work less on their own in private practice, as more and more of them began working as employees in large bureaucratic organizations, for example, companies that built automobiles, airplanes, and radio and television systems. Engineers working in large corporations or government agencies have less autonomy; they report to individuals who are higher up in the organizational hierarchy and are not necessarily engineers.

A common way of thinking about twentieth-century engineers has been to see them as caught in a tension between their commitments as engineers and the demands of their employers, typically the demands of the business world. As professionals, engineers are expected to make decisions based on their expertise, and they are expected to adhere to standards set by their profession. However, at the same time they are expected to do what their employers want, which may require considering riskier alternatives, alternatives that don't quite square with the time the engineer would need to ensure safety or that cost less than is needed to achieve a certain safety level. Although the demands of an employer and the standards of professional behavior are not necessarily or always in conflict, it has been widely recognized that they do at times conflict.

Because engineers of the twentieth century had both professional commitments and obligations to their employers, they faced difficult dilemmas. The classic dilemma, which received a good deal of attention (and still does), is that of whistleblowing. In such situations, the engineer is forced to decide whether to blow the whistle on her or his employer because the employer is disregarding the engineer's concerns about the safety of a product about to be released or already in use. The engineer is torn between his or her professional judgment and pressure from a supervisor to disregard the issue, for example, because the company is under financial pressure or pressure to meet a deadline. Chapter 5 is devoted to the topic of whistleblowing.

For now, the important thing to note is that this tension—the tension between being a professional and being an employee—persists in the twenty-first century and has everything to do with what it means to be a member of a profession: a professional. Although the distinction between professionals and mere employees is drawn less frequently and less starkly today, it is not because the situation of engineers has changed. Rather, it is because so many of the traditional professions have changed in ways that make them look like twentieth-century engineers. Consider, for example, doctors, who in the nineteenth and first half of the twentieth century predominantly had individual private practices, while today they mostly work in large bureaucratic contexts, such as in large private practices, in large public institutions, and in hospitals. Even if they work individually or in a small practice, much of their activity is controlled by insurance companies and government agencies. This means that doctors now have a tension that parallels that of engineers; they are caught between pressures to do what their professional expertise tells them and the pressures to do what their employers want them to do or what the regulations require them to do. In short, the tension between commitments as professionals and commitments to employers is commonplace in most professions today.

Although the notion of profession is now more commonly associated simply with college degrees, white-collar work, high salaries, and social status, sociologists continue to try to pin down the distinguishing features of professions. For example, in 2011, Elizabeth Gorman and Rebecca Sandefur used four characteristics to distinguish professions: (1) expert knowledge; (2) technical autonomy; (3) a normative orientation toward service to others; and (4) high status, income, and other rewards.[5] Engineers of the twenty-first century seem to fit these criteria. However, the profession is complex, so we need to consider to what extent and how engineering embodies these characteristics.

Being an engineer is unlike many other jobs insofar as it requires the use of expert knowledge. Engineers acquire expert knowledge through undergraduate education and hone their expertise with work experience. ABET's development of curricular criteria that all engineering programs must meet attests to the existence of a body of knowledge that all engineers must master before they can practice effectively. Expert knowledge is the most fundamental claim that engineering can make regarding its status as a profession. Because of their special knowledge, engineers are able to understand what lay people cannot; because of their expert knowledge, engineers are put in positions to make decisions on technical matters about which nonengineers are ignorant.

Expert knowledge leads to the second characteristic of professions. The expert knowledge of engineers provides the basis for a claim to technical autonomy. Technical autonomy is different from the kind of autonomy engineers might have had in the nineteenth century in private practice. Nevertheless, technical autonomy is significant. Even when they work in large organizations, engineers are given assignments that allow them to use their technical expertise to make judgments for their employers. Groups of engineers are also asked to set standards in various industries. Again, the power and authority (autonomy) that engineers have in these contexts derives from their technical expertise.

Engineering also has a normative orientation. A normative orientation is one that includes norms of good and bad. Professions with a normative orientation have standards designating some actions as good, desirable, obligatory, or permissible and others as bad, undesirable, or impermissible. Engineering codes of ethics provide this normative orientation by specifying the obligations, values, and constraints with which engineers must operate. Professions (unlike other occupations) generally make a commitment to service, and the engineering codes make such a commitment when they specify that engineers are to hold the health, safety, and welfare of society paramount and are to be faithful to their clients. As Gorman and Sandefur explain, the service orientation has two aspects. First, professionals are expected to put their clients' interests above their own; they are expected not to take advantage of their clients' lack of knowledge. Second, professionals are expected to serve the public good. Both commitments are found in engineering codes of ethics. Last, engineers have the high status, income, and rewards associated with professions. It is important to note that high status and income result from the other three characteristics. That is, the justification for high status and high salaries is that engineers have special expertise and technical autonomy and that they make a commitment not just to pursue their own self-interest but to serve their clients, protect the public, and uphold certain standards of behavior.

Recognizing these four characteristics as essential to establishing any occupation including engineering as a profession provides the context for understanding the importance of codes of ethics. Codes of ethics are an essential part of constituting a group as a profession. Before we delve more deeply into this idea, however, a brief sidetrack will clarify who is included in this discussion of engineering.

PROFESSIONAL ENGINEERS

In thinking about engineering as a profession, confusion may arise between engineering in general and professional engineers (PEs).[6] PEs are a special subset of engineers. Professional engineers are required to obtain a license, and the license gives them special rights and responsibilities. For example, only a licensed engineer can "prepare, sign and seal, and submit engineering plans and drawings to a public authority for approval, or seal engineering work for public and private clients." A PE is often required to be in "responsible charge" of firms that provide certain kinds of engineering services, and only PEs can, in certain contexts, serve as qualified expert witnesses.[7]

Engineers become PEs through a rigorous process. As explained on the website of the NSPE, in order to become licensed, engineers must "complete a four-year college degree, work under a Professional Engineer for at least four years, pass two intensive competency exams and earn a license from their state's licensure board. Then, to retain their licenses, PEs must continually maintain and improve their skills throughout their careers."[8]

Although the NSPE is devoted to providing resources and support to PEs and potential PEs, it does not issue licenses. Only state boards do this. The National Council of Examiners for Engineering and Surveying, a national nonprofit organization, oversees and provides support to state licensure boards.

The special authority of PEs to sign and seal drawings and be responsible for certain kinds of work entails liability. PEs can be sued for incompetent behavior, and they can have their state licenses revoked by state boards. So, state licensure boards can enforce standards. They can discipline PEs for violations of the state board's code of ethics. Disciplinary actions include penalties as well as revocation of licenses. In light of the discussion in chapter 1, it is worth noting that state licensure boards sometimes require violators to take an ethics course.

Although the percentage of engineers who seek licensure is difficult to determine, it is generally given as 20 percent. That is, 20 percent of engineers in the United States are licensed. The NSPE estimates that there are more than two million practicing engineers in the United States.[9]

Because the bulk of engineers practicing in the United States and elsewhere are not PEs, the primary focus of this book is on all engineers and not just PEs. There will be a few points at which the relevance of licensure will come into play, but for the most part the issues discussed apply to PEs and non-PEs alike.

DO ENGINEERS NEED CODES OF ETHICS?

In order to understand the role of codes of ethics in engineering, the first thing to note is that they are not a standalone feature (of engineering or any other profession). They are one part—albeit an important part—of the strategy by which engineering has established itself as a group worthy, not just of the title "profession," but of some degree of autonomy and public trust. The mere fact that the profession has codes of ethics indicates that hiring an engineer is better than hiring someone who is not an engineer. Of course, the educational requirements for engineering are also a major factor in this claim. The fact that engineers have spent four years mastering a body of knowledge—a body of knowledge that is continuously being improved on and updated—goes a long way toward distinguishing engineers from others who might do similar work. However, special expertise alone is not sufficient. To be worthy of public trust, engineers must demonstrate that they have a commitment to using the special knowledge that they have in ways that serve the public or, at least, in ways that do not harm the public. Codes of ethics proclaim that commitment.

This takes us back to a criticism mentioned earlier. Skeptics disparage codes of ethics for being public relations tools; they claim that codes promote the reputation of engineers and little else. Putting codes of ethics in the context of professionalization counters this criticism by showing that although codes do in fact promote the public image and reputation of engineers, this is no trivial matter. Public image has to do with public trust. Public trust is essential for engineering to exist as a profession. It is essential to establishing engineers as a distinctive group different from others without degrees or with degrees in other fields of higher education. Public trust is essential to engineers receiving some degree of autonomy so that they can use their special expertise.

Of course, the skeptics might insist that engineers put forth codes of ethics only so that they appear to be worthy of trust. For this reason, it is important to recognize that in order to demonstrate that engineering is a profession worthy of public trust, engineers must not only produce and publish a code of ethics but must follow it. Once a code has been promulgated, engineers must show that the code has significance for them, that it affects their behavior, and, most important, that they adhere to it. Engineering professional organizations sometimes engage in activities that reinforce their codes of ethics, for example, by discussing ethical issues in professional society journals or giving awards to engineers who have acted in morally admirable ways. These activities encourage members to take the code of ethics seriously and to make it part of their understanding of their professional responsibilities. In this respect, codes of ethics help to define what it means to be an engineer and what it means to hire an engineer.

Insofar as codes of ethics are understood as part of establishing that a profession is worthy of public trust, it is tempting to think that they are directed primarily at the public (and engineers themselves). Yet the public is only one of the audiences to which codes are directed. Codes of ethics also play an important role in relation to employers. In presenting the history of the ASCE's code of ethics, Sarah Pfatteicher notes that the ASCE rejected the idea of adopting a code of ethics for many years. At the time, engineers were aware of the importance of reputation and sought to enhance their reputations through membership in ASCE, but they resisted creating a code of ethics in part because they feared that such a code would diminish rather than enhance their autonomy. That is, some believed that a code of ethics would constrain them by specifying how to behave. In the discussion leading up to the decision to adopt a code of ethics (in 1914), image was unabashedly one of the purposes for adopting a code, but it wasn't just public image. Pfatteicher explains that the early codes of ethics adopted by ASCE were intended to "document and publicize existing standards of behavior (largely for the benefit of potential employers)." So, the original ASCE code was targeted to set the expectations of those who would employ engineers.[10]

In this respect, codes of ethics can be useful in supporting engineers who disagree with their employers. Mike Martin and Roland Schinzinger explain it this way: "A publicly proclaimed code allows an engineer, under pressure to act unethically, to say: 'I am bound by the code of ethics of my profession, which states that . . .' This by itself gives engineers some backing in taking stands on moral issues. Moreover, codes can potentially serve as legal support for engineers criticized for living up to work-related professional obligations."[11]

In thinking about codes of ethics, then, it is important to remember that they are directed at multiple audiences, including the public, employers, and members of the profession themselves (engineers). Engineers have relationships with others such as contractors, inspectors, and financiers, and these individuals are also informed by codes of ethics about what to expect from engineers.

Some of the criticisms of codes of ethics seem to presume that codes are directed exclusively at engineers and at controlling their behavior, as if codes of ethics were like laws. They are then criticized for being ineffective at regulating behavior because they don't have enforcement power. Here it is important to remember the difference between PEs and other engineers. The codes of ethics promulgated by state licensure boards for PEs do have enforcement power. PEs can be punished for failure to adhere to these codes. On the other hand, codes of professional societies are susceptible to this criticism. They do not have enforcement power. However, this seems a weak criticism because codes of ethics are not exactly intended to coerce behavior. They have, as we have seen, other purposes. Moreover, professional societies can and do encourage members to adhere to the codes; they just don't punish violations. Instead, they promulgate codes of ethics in order to inform, encourage, and inspire engineers to behave in particular ways.

One argument against codes of ethics posits that codes of ethics are not merely window dressing but actually counterproductive. They can undermine ethical behavior by leading to complacency. The existence of codes of ethics, the argument goes, suggests that one need only follow the rules to be ethical. The codes set out what might be thought of as a minimum standard. The criticism is that individual engineers may, then, do only the minimum. They may engage in unethical behavior that isn't strictly covered by the codes. Along these lines are seemingly noble arguments suggesting that the existence of codes of ethics implies that ethics is simply a matter of following rules when, in reality, ethical decision-making often involves discretion and judgment.[12] The concern here is that engineers will believe that if they follow the rules, they have done all they need to do.

The problem with this criticism is that the statements in most professional engineering codes are too general to be treated or used merely as rules to be followed. A code of ethics cannot anticipate all the situations in which engineers may find themselves, so codes consist of broad and general principles. Even if engineers wanted to follow rules, codes of ethics don't make it easy. The broad, general statements in the codes must be interpreted and applied to the details of a specific situation. Since engineers still must use discretion and judgment in applying the statements in codes of ethics, it is hard to see how they could make engineers complacent.

Related to the complacency criticism is the criticism, mentioned earlier, that codes are ineffective because they lack enforcement power. This criticism, like the complacency criticism, takes codes of ethics to be a set of rules that engineers are expected not to break. As the argument goes, if the rule breakers are not penalized, then the rules will not be taken seriously.

The trouble with this critique is, as already suggested, that most codes of ethics tend to consist of general principles that point individuals in the right direction but do not tell them exactly what to do. Take, for example, one of the edicts about honesty. The NSPE code states that its members should "avoid deceptive acts," and the IEEE Code of Ethics specifies that members should "be honest and realistic in stating claims or estimates based on available data." Although these rules may seem simple and straightforward, in fact, when it comes to providing information to a client or the public, they don't provide the kind of guidance needed. Of course, engineers should not outright lie or falsify documents. On the other hand, there are more complicated situations in which how to be honest or avoid deception is not obvious or straightforward. For example, being honest with a client can't mean telling them every technical detail of what you will do since the client may not be an engineer with the technical expertise to understand. Not telling a client about the intricate details of a project is not being dishonest, but it is not easy to figure out what to tell and what is okay not to tell. This is all to say that professional codes of ethics are not intended to be specific enough to be easily enforced.

When it comes to the purpose of codes of ethics in relation to practicing engineers, instead of thinking of codes as sets of rules to be enforced, it is better to think of them as representing the collective wisdom of members of the profession. Typically, codes of ethics are developed (and changed) through a process involving many engineers. Drafts are made, comments are solicited from a wide range of members, debate and discussion is encouraged, and then agreement is sought. In this process, engineers express what they have learned and put this into a relatively small set of statements describing the principles that engineers should embrace. In this way, codes of ethics tell individual engineers, and especially new members, what experienced members of their profession believe are the key principles to be used in guiding behavior.

Codes of ethics can be especially informative for early career engineers. The codes help to socialize new members into the profession. They provide general guidance as to how to behave. Often new engineers learn how to conduct themselves in their first jobs. If their first job happens to be in a place in which standards are sloppy and unprofessional, the new engineer may come to believe that this is what is expected of engineers. On the other hand, if the first job is one in which the culture and practices are highly professional and ethical, the new engineer may learn that this is how engineers behave. The problem is the randomness of first job experiences. Codes of ethics counterbalance this randomness by telling all individuals entering the field what is expected of them as engineers regardless of where they work.

So, codes of ethics are directed at multiple audiences and they serve different purposes in relation to each. Codes lay out the commitments of engineers informing employers and the public about what to expect from engineers. They help to establish the image and reputation of engineers, and they promote trust in the profession and in individual practitioners. They shape the expectations of all those who interact with engineers, including employers, clients, contractors, government officials, the media, and others. In this respect codes of ethics are part of the strategy by which engineering constitutes itself as a profession. When it comes to engineers themselves, codes of ethics represent the collective wisdom of members and provide general guidance to individual members, but they cannot necessarily provide the specific guidance an engineer in a particular situation might want. As the collective wisdom of the members of the profession, codes of ethics provide vital information for the socialization of new engineers into the profession.

OTHER DEBATABLE ISSUES INVOLVING CODES OF ETHICS

In addition to debate about the need for codes of ethics, several features of codes continue to be debated. One debate concerns what should be included in a code of ethics. Some argue that codes of ethics should not include prohibitions on behavior that is illegal. For example, some codes have a prohibition against bribery though there are laws against bribery. The argument is that it goes without saying that engineers should not engage in illegal activity, so there is no need to prohibit bribery in a code of ethics. The counterargument is that certain issues can arise in engineering practice that are especially problematic for the profession. These matters—even though regulated by law—should be emphasized in codes of ethics. Another focus of debate as to whether codes should prohibit illegal behavior centers on statements against discrimination—prohibitions against discrimination on the basis of race, gender, religion, or sexual orientation. Some argue that engineering codes of ethics should include clauses of this kind. Others argue that such clauses are unnecessary because such behavior is illegal.

Another debate on the content of engineering codes has to do with sustainability. Diane Michelfelder and Sharon Jones argue that sustainability should be included in engineering codes of ethics not just as another statement but as part of the paramountcy clause—that is, the clause should specify that engineers should hold paramount sustainability as well as the safety, health, and welfare of the public. Their argument is intriguing because it is based on the idea that sustainability includes social justice. Activities directed at achieving sustainability have distributive effects—that is, they affect different groups of people differently. This is true inter- and intragenerationally. Making social justice part of sustainability implies that achieving sustainability should promote, and not be at the cost of, social justice. Hence, inclusion of sustainability in engineering codes of ethics will promote engineering that contributes to social justice, a public good.[13]

Debates about what should be included in engineering codes of ethics take us back to the multiple audiences and purposes of codes of ethics. Even though some forms of behavior are illegal, it might be important for an organization of engineers to emphasize to their members or to the public or to employers that engineers are committed to avoiding such behavior. For example, in the case of bribery there may be some value in emphasizing that for engineers, bribery is not only illegal but also contrary to their professional commitments, that engaging in such behavior is unprofessional as well as illegal.

Another matter of debate is whether a code of ethics applies to all engineers with a degree in the field or only engineers who are members of a particular professional organization. Suppose you have a degree in mechanical engineering and a job that is specifically for mechanical engineers but you are not a member of the ASME. Does the ASME Code of Ethics apply to you? Mechanical engineers are not required to belong to the ASME. One argument for insisting that the codes apply to all engineers who have the degree and work in the field regardless of whether they are members of the professional association is that it seems unfair that some engineers would benefit from the existence of codes of ethics while not bearing the burden of restraining their behavior in accordance with the code. All members of an engineering field benefit from the existence of the code of ethics in that field because the code helps to establish the field as a profession worthy of public trust, some degree of autonomy, and high salaries and status. Codes of ethics promote confidence and trust in the engineers in that field. Hence, those who don't belong to the professional organization and especially those who act contrary to the code are benefiting without having to bear the burden of operating within the constraints of a code. They should, therefore, be expected to adhere to the code.

Last, as mentioned at the start of this chapter, there are many engineering codes of ethics and professional conduct. Charles

Harris has suggested that it would be better to have one unifying code for all engineers, and not just engineers in the United States but for engineers globally.[14] On the one hand, a single unifying code seems plausible since the existing engineering codes of ethics are all similar. Indeed, a code of ethics that applied globally would help engineers who practice in a variety of countries. A global or universal code would eliminate confusion and perhaps raise the standards for engineering across the world. On the other hand, although a single code might be a good thing, given the process by which codes of ethics are developed and adopted, it would be no small feat to develop a single code that engineers in every field and every country could agree on.

These debates illustrate that codes of ethics are an ongoing topic of discussion among engineers. Codes evolve and change over time. They are living documents that both represent and reflect how engineers understand their profession and its role in the world.

CONCLUSION

Does engineering need a code of ethics? A multitude of engineering organizations in the United States and worldwide have adopted codes of ethics and professional conduct. Yet, as has been demonstrated, the need for or wisdom of having such codes can be challenged. Criticisms vary from claiming that codes of ethics are merely public relations tools to the assertion that they lead to complacency to the charge that they are ineffective because they lack enforcement power. Although these critiques are worthy of attention, in order to appreciate the value of codes of ethics, it is essential to understand them as part of a strategy to constitute engineering as a profession. In this respect, codes of ethics cannot be understood in isolation. They are one part of a broader strategy that engineering has adopted to establish its role in the world and to set expectations for engineers themselves, employers, the public, and others. In addition to putting codes of ethics in the broader context of professionalization, codes of ethics should not be understood as a set of rules to be followed blindly. They are not meant to be the final word on ethical behavior in engineering. They represent the collective wisdom of engineers in a form that can broadly guide members and help to socialize new members. Moreover, codes of ethics can inspire engineers to behave well and to exhibit the kind of moral courage that was discussed in chapter 1.

SUGGESTED FURTHER READING

Layton Jr., Edwin T. *The Revolt of the Engineers: Social Responsibility and the American Engineering Profession*. Baltimore: Johns Hopkins University Press, 1986.

Pfatteicher, Sarah K. A. "Depending on Character: ASCE Shapes Its First Code of Ethics." *Journal of Professional Issues in Engineering Education and Practice* 129, no. 1 (2003): 21–31.

Reynolds, Terry S., ed. *The Engineer in America: A Historical Anthology from Technology and Culture*. Chicago: University of Chicago Press, 1991.

3 HOW SHOULD ENGINEERS THINK ABOUT ETHICS?

EVERY person experiences ethical dilemmas both large and small: How should I treat my dying grandfather? Should I give money to the homeless beggar? Should I use information that I am not supposed to have in preparing my homework or for an exam? In chapter 2 we established that whatever ethical challenges engineers face *as persons,* they also have distinct ethical obligations and commitments *as engineers.* Codes of ethics and professional conduct suggest that engineering ethics is a matter of professional ethics. Although this suggestion is right—that is, engineers should think of engineering ethics as involving their responsibilities as professionals—there is more to engineering ethics than adherence to a professional code of conduct. Not only the scope of engineering ethics but the relationship between engineering ethics and ethics more generally needs further exploration.

In this chapter, we consider several ways of thinking about ethics. First, the notion of professional ethics is expanded beyond codes of conduct to embrace a deeper connection between engineering and society. Engineers have a broad responsibility to consider the social implications of their work. Second, concepts and frameworks from theoretical ethics are introduced as systematic ways of thinking about ethics. Ethical concepts and theories are relevant broadly, not just to engineering, and they are put forward here as ways of thinking about ethics that are relevant to personal matters as well as to professional decision-making and broader social and policy issues. Last, in this chapter, some attention is given to practical ethics—that is, to how engineers should figure out and decide what to do in real-world situations. Here the suggestion is that ethical decision-making is not just a matter of applying codes of ethics or ethical theories as if they were algorithms but is much more like the activity of designing solutions that fit the context. In real-world situations, one must design courses of action that take a variety of considerations into account, and this often involves making difficult trade-offs.

Although this chapter has an overarching question—how should engineers think about ethics?—it is not framed as a single debate. Professional ethics, theoretical ethics, and practical ethics are not competing perspectives: they are three different ways to think about the ethical challenges that engineers face. Within each perspective there are, however, many debatable issues. For example, in discussing professional ethics, although it is broadly agreed that engineers have some responsibility for the social consequences of their work, there are disputes about what is encompassed in this responsibility and the lengths to which an engineer should be expected to go to fulfill it. As well, in discussing ethical theories, the adequacy of each theory can be debated. And when it comes to practical ethics, the idea that ethical decision-making is a matter of deducing answers from ethical codes or principles or an open-ended process similar to solving design problems can be debated.

AN EXPANDED NOTION OF PROFESSIONAL ETHICS

As suggested, one way to think about engineering ethics is as professional ethics. Using this approach, engineering ethics is focused on how individual engineers and engineers collectively should behave—what they should care about, what they should refrain from doing, and what they must do. Whistleblowing, a classic topic in engineering ethics, exemplifies this. Whistleblowing typically involves a decision to go against the wishes of a client or an employer; this seems to violate the obligation to act faithfully on the client or employer's behalf in order to fulfill the obligation to protect the public. So, there is nothing wrong with saying that engineering ethics is a matter of professional ethics. What would be wrong is to say that engineering ethics is merely or exclusively a matter of fulfilling one's professional duties as specified in professional codes of conduct. This is much too narrow a conception of engineering ethics insofar as it neglects the deep connection between engineering and people—engineering and society—which is the basis for many of the principles in the codes of ethics.

The connection between engineering and people is not just a connection between engineering and individuals. Engineering affects the lives of individuals, but it also affects the quality and nature of the societies people live in and the distribution of benefits and burdens among individuals and groups. A few decades ago, many engineers rejected the idea that there was any significant connection between engineering and ethics. Claims like the following were commonplace:

> "The job of engineers is to figure out how nature works and to manipulate natural phenomena in order to develop useful products and processes."
> "The engineer's job is to get the technical right; it is not the engineer's job to worry about society."
> "Engineers provide technological solutions, and others decide whether the solutions should be used."
> "Technology is neutral; values come in only in the way people use it."
> "Since engineering is about machines and devices, it has nothing to do with ethics."

Such claims are rarely made today and, if made, are generally rejected. They are worth mentioning here because they reflect a deep but false presumption that technical and social matters are separate. Today, engineers as well as the public readily recognize that engineering, technology, and society are intimately intertwined. Changes in one can result from and affect the other.

There are at least two important ways in which the intertwining needs to be acknowledged in thinking about engineering ethics. First, engineering is a social activity. It involves people working with one another to accomplish tasks; it involves human beings organized in distinctive ways, for example, in teams and organizational hierarchies. As well, engineering is influenced by the world around it. The history of engineering shows that engineering has changed over the years not just because of technological developments but also because of social ideas and forces that affected the organization of the profession and how engineers work. The emergence of professional societies and the development of licensing procedures are two examples of social organizational change that have affected engineering practice.

Second, engineering and technology powerfully affect the way people live. Engineering and technology affect the quality and character of human living and human experience. They affect what people do and how, where, and when they do it. Bridges and highways affect where people go; cell phones have changed who interacts with whom, when, and how often; biomedical devices affect who lives and dies. Whether one considers global relations enabled by a vast global communication network or our most intimate personal relations affected, for example, by reproductive technologies, in the twenty-first century it is hard to imagine any dimension of human life that has not been influenced by technology and engineering.

So, engineering/technology and society are intimately intertwined. In fact, they are so tightly connected that the phrase "technology affects society" does not adequately capture the connection. For one thing, the influence is bidirectional. Engineering and technology influence the character of society *and* society influences the character of engineering and technology. For another, technology and human behavior work together to make technologies work. Consider large-scale technological endeavors such as nuclear power plants or Facebook. We may refer to these as technological, but their operation depends both on machines and devices *and* on humans behaving in certain ways. In the case of Facebook, for example, you may think it consists primarily of software, a graphical interface, computers, and monitors, but in order for Facebook to function, many people have to go to work each day to maintain the operation of the system. It is not just those who maintain the computer servers; for Facebook to work, a complex organization of people is essential, including top-level administrators, administrative assistants, people who manage finances, human resource officers, building maintenance workers, and, of course, users.

Once the deep connection between engineering and people/ society is acknowledged, it is a small step to ethics. Ethics has to do with how human beings treat one another, and since engineering/technology affect the ways in which people organize themselves and behave toward one another, engineering is an inherently ethical endeavor. Ethical questions can be raised about the nature of the social relationships involved in engineering and the nature of the effects on people that engineering produces. This means that engineering ethics has to do with how engineers conduct themselves in their work *and* with how their behavior and their activities as engineers affect people and social arrangements. It is worth mentioning here that the work of engineers also affects the natural environment and nonhuman living creatures; this, too, is part of engineering ethics.

The Scope of Responsibility for Social Consequences

Although the connection between engineering/technology and people/society is undeniable, a skeptic might challenge the implications of this connection for the social responsibility of individual engineers. That is, a skeptic might claim that individual engineers should not be held responsible for the social implications of their work. In defending this claim, the skeptic might make several points. First, engineers are not the only actors making decisions that affect the character of human societies. In fact, engineers are not even the only actors determining what technologies are developed and how they are used. Many other actors are involved—entrepreneurs, financiers, regulators, marketers, media, environmental groups, and consumers, to name a few.

Second, engineers are generally distant from the social effects of their work. Often they can't see how their work will be used, let alone see the effects it will have. For example, an engineer working on the development of a biomedical device may work primarily in a laboratory and rarely see the people whose lives will be significantly improved if the device is successfully developed. In this case, the connection to people is closer and the effect positive. In contrast, engineers working on such things as transistors or turbines or new materials may have little idea how their work will ultimately get used in a product or process or how it will be put together with other parts or processes. Hence, it is difficult to see how their work will affect what people do.

Third, engineers often work on small parts of large endeavors, so they cannot foresee how the whole to which they contribute will turn out. This means, again, that engineers cannot foresee the social implications of their work. Engineers may be genuinely uncertain as to how their work is going to be used and therefore uncertain as to its effects on individuals or society.

The points that the skeptic makes here are not wrong. In fact, they illustrate the complexity of engineering practice and the challenge of being an ethical engineer. Yes, engineers work with others so that the effects of their efforts are a combination of their effort and that of others. Yes, the effects of engineers' work are often distant, so it takes some effort and even imagination to foresee the consequences. And, yes, engineers often work on only a small component of a larger endeavor, so, again, it is difficult to see what the consequences will be. All of this is true and yet none of it justifies letting engineers off the hook of responsibility. Indeed, it would be dangerous to absolve engineers of responsibility, for to do so is to suggest that engineers should work blindly. It is to suggest that they should be slaves to their bosses or employers—just do what they are told without asking questions—and, perhaps more important, that they should cease to be moral beings when they work. In a sense, such a view frames engineers as machines; they are to work but have no thoughts about what they are doing.

Rather than absolving engineers of responsibility, the points that the skeptics make provide a strong basis for just the opposite conclusion. Because of the complexity of their relationship to the effects of their work, engineers have a responsibility to make a special effort to keep in mind the connection between their work and the people who will be affected by it. Engineers should think about and inquire about the contexts in which they are working. No matter whom they are working for or on what project they are working, there are always some ideas about how the work will be used, whom it will serve, and what its benefits will be. Generally, projects are pursued precisely because someone has ideas about what the project will accomplish, whom it will help, and how it will help. Engineers generally don't and shouldn't ignore these ideas.

And, of course, many engineers work on projects or do jobs in which they are not distant from the people affected and the social implications. They know who is likely to be affected and how. For example, an engineer may know the neighborhood in which a building is being constructed, the industry in which a mechanical device or process will be used, the types of consumers that are targeted to benefit, and so on. In many cases, engineers not only know who will be affected but also have some idea of the risks and which groups of people are likely to bear those risks.

So, there are compelling reasons not to let engineers off the hook of responsibility for the work that they do. This responsibility can be acted on in a wide range of situations, for example, when engineers decide what job to take or what project to work on or when they decide whether to pursue a suspicion they have about a potential danger or risk or whether to report wrongdoing. Even though the social implications of an individual engineer's work may be distant and difficult to anticipate, the connection between engineering and society is always there, and acknowledging the connection is essential to understanding the social significance of what one is doing. Only when engineers understand this connection can they fully understand engineering ethics.

We will return to this topic in chapter 8 when we explore engineers' responsibilities for social justice. For now, one important caveat is worth noting. The focus in the preceding discussion has been on individual engineers, and the story is quite different when it comes to engineers acting collectively. Collectively engineers have more power to address the social implications of large engineering endeavors such as transportation systems or energy systems and their safety and their effects on health and the environment. So, the connection between engineering and society has implications for engineers collectively as well as for individual engineers.

THEORETICAL ETHICS

Setting aside the "engineering" part of engineering ethics, there is a long tradition of ethical theory and ethical thought that is relevant to engineers' thinking about ethics. In its deepest sense, ethics has to do with how human beings should treat one another, right and wrong, good and bad, what we owe one another, and what is just or fair. As such it is an inherently normative domain; it focuses on what should happen, not what does happen. For these normative matters, moral philosophy provides a systematic body of thought. For centuries, moral philosophers (Western and Eastern alike) have sought to understand the foundations—the underpinnings—of morality. In the past several decades a substantial amount of attention in engineering ethics has been focused on three major theories: utilitarianism, Kantian ethics, and virtue ethics.

These three theories are introduced here as examples of theoretical, systematic thinking about ethics. They are not the be all or end all of moral philosophy and should not be taken as such. They are introduced here as starting places for thinking about ethics theoretically. Each of the theories provides a framework, a set of concepts, and language that are useful for discussion and understanding of ethical issues and dilemmas. Each theory illustrates a way in which ethical analysis can bring better understanding.

Utilitarianism

Utilitarianism was developed by philosopher Jeremy Bentham (1748–1832) and elaborated and extended by John Stuart Mill (1806–1873). According to utilitarianism, what makes behavior right or wrong depends on the consequences. For this reason, utilitarianism is also often referred to as a form of consequentialism. What is important according to consequentialists is the outcome or results of behavior and not the intention of the person acting. Traditional utilitarianism is focused on a particular type of consequence—namely, those that produce happiness or well-being. Actions are good when they produce the most happiness and bad when they produce net unhappiness. This philosophy is called utilitarianism because it holds that actions, rules, or policies are good because of their utility in bringing about good consequences.

The fundamental principle of utilitarianism is this: Everyone ought to act so as to bring about the greatest amount of happiness. Utilitarianism recommends that when you are deciding what to do, you should assess the consequences of your acting in one way or another and take the action that will bring about the most overall good consequences.

Why focus on consequences and neglect intentions? Utilitarians argue that happiness is the only thing that is so valuable that it requires no justification. In utilitarianism a distinction is made between things that are valued because they lead to something else (instrumental goods) and things that are valued in themselves—that is, for their own sake (intrinsic goods). Money is a classic example of something that is instrumentally good. It is valuable, not for its own sake, but rather as a means for acquiring other things. On the other hand, intrinsic goods are not valued because they are a means to something else. They are valuable in themselves. Happiness is an intrinsic good. Friendship is also often used as an example of another intrinsic good. We don't value friends because of what they can do for us; we value them in themselves.

According to utilitarians, a moral theory must be based on something that is so valuable that people should commit their lives to it. Happiness, they argue, is the only good that can serve in this role because happiness is the only thing that is truly valued for its own sake. Indeed, some utilitarians claim that all other goods are desired as means to happiness. Even friendship, health, and knowledge, they would say, are valued because they lead to happiness.

With happiness as the ultimate good, then right and wrong follow; the right action is the one that brings about the most net happiness, and the wrong action is the opposite. When a person faces a decision, the person should consider possible courses of action, anticipate the consequences of each alternative, and choose that action which brings about the most happiness—that is, the most overall net happiness. To be sure, the right action may be one that brings about some negative consequences (unhappiness), but that action is justified if it has the most net happiness or the least net unhappiness of all the alternatives.

Utilitarianism should not be confused with egoism. Egoism is a theory that specifies that one should act so as to bring about the greatest happiness or good consequences *for one's self*. What is good is what makes me happy or gets me what I want. Utilitarianism does not say that one should maximize one's own good. The sum total of happiness in the world is what is at issue. Thus, when you evaluate your alternatives, you have to ask about their effects on the happiness of everyone. This includes effects on you, but your happiness counts the same as the happiness of others. It may turn out to be right for you to do something that will diminish your own happiness because it will bring about a marked increase in overall happiness.

As a general theory, utilitarianism has been influential in such fields as economics and policy analysis. Legislators and public policy makers seek policies that will produce good consequences, and they often opt for policies that may have some negative consequences, but they do so with the idea that on balance there will be more good consequences than bad. Cost-benefit and risk-benefit analysis are rooted in utilitarian thinking insofar as they specify that one should balance risks and benefits against one another in making a decision.

There are many forms of utilitarianism, and one contentious issue is whether the focus should be on *acts* or *rules*. Some utilitarians have recognized that it would be counter to overall happiness if each of us had to calculate what the consequences of every one of our actions would be before we act. Not only is this impractical—because it is time consuming and because sometimes we must act quickly—but often the consequences are impossible to foresee. There is thus a need for general rules to guide our actions in ordinary situations. Accordingly, some argue that we should focus on rules rather than acts.

Rule utilitarians argue that we should adopt rules that, if followed by everyone, would, in the long run, maximize happiness. Take, for example, telling the truth. If individuals regularly told lies, it would be very disruptive. You would never know when to believe what you were told. In the long run, a rule obligating people to tell the truth has enormous beneficial consequences. Thus, even though there are sometimes occasions when more good will come from lying, the rule "Tell the truth" is justified on utilitarianism. "Keep your promises" and "Don't reward behavior that causes pain to others" are also rules that can be justified on utilitarian grounds. According to rule utilitarianism, if it is true that when people follow the rule, long-term net good consequences result, then the rule is justified and individuals ought to follow the rule.

Act utilitarians believe that each action should be considered on its own. They believe that even though it may be difficult for us to anticipate the consequences of our actions, that is what we should try to do. Rules can be used, but they are not strict; they should be understood to be rules of thumb—that is, general guidelines that can be abandoned in situations where it is clear that more good consequences will result from breaking them. Rule utilitarians, on the other hand, take rules to be strict. People should follow the rules even if in one particular case, the negative consequences outweigh the benefits. Rule utilitarians worry that breaking the rules on occasion will undermine the effect of rules and will therefore have overall net negative consequences.

Utilitarianism can be used for engineering ethics in many ways. An engineer might use the theory to think through a situation where the right thing to do is not so obvious. The engineer would identify the positive and negative effects of one course of action or another and calculate which would be better. The theory might also be used to understand the justification for rules or practices such as those specified in the codes of ethics. A rule—such as the one specifying that engineers should avoid deceptive acts—is justified on grounds that overall good consequences result from engineers following it. And good consequences here does not mean consequences only for engineers, it means consequences for engineering and engineers as well as clients, employers, and the public. Another related way that the theory might be used is to make the case for eliminating or changing a professional rule or practice. Suppose, for example, that policy makers want to change an existing practice by requiring the signature of a licensed engineer for an activity that previously could be done without it. A change of this kind might be justified in a utilitarian framework—that is, on grounds that it would reduce the risk of harm, thereby improving the positive effects of the activity.

Critique of Utilitarianism

Utilitarianism is an important framework for thinking about ethics, but it is not without problems. Some have suggested that it does not adequately capture what is important about ethical thinking. Perhaps the most significant criticism is that the theory goes against strongly held moral intuitions insofar as it seems to tolerate, if not recommend, that burdens be imposed on some individuals for the sake of overall good. Although it is true that every person is to be counted equally—that is, no one person's happiness or unhappiness is to be counted more than another's—there are situations in which overall happiness would result from sacrificing the happiness of a few. Suppose, for example, that having a small number of slaves would create great happiness for a large number of individuals. The individuals who were made slaves would be unhappy, but this would be counterbalanced by significant increases in the happiness of many others. Such a situation would seem to be justifiable in a utilitarian framework. Another more contemporary example is to imagine a situation in which by killing one person and using their organs for transplantation, ten lives could be saved. Killing one to save ten would seem to maximize good consequences.

Critics of utilitarianism argue, then, that since utilitarianism justifies practices such as slavery or more broadly sacrificing some for the sake of others, it is not a cogent ethical theory. In their defense, utilitarians point to the difference between short-term and long-term consequences. Utilitarianism is concerned with all consequences, and when long-term consequences are taken into account, such practices as slavery and killing innocent people to save other lives cannot be justified. In the long run, such practices have the effect of creating so much fear in people that net happiness is diminished rather than increased. Imagine the fear and anxiety that would prevail in a society in which anyone might at any time be taken as a slave. Or imagine the reluctance of anyone to go to a hospital if there was even a remote possibility that they might be killed if they happen to be at the hospital at a time when a major accident occurred and organs were needed to save many victims. This defense shows that utilitarianism is not a simple theory—that is, calculating the consequences of particular actions and practices is not as easy as it might at first appear.

Nevertheless, and despite the utilitarian defense in terms of long-term consequences, the concern of critics is that, at its heart, utilitarianism frames individuals as valuable in terms of their social value—that is, the value of a person has to do with the person's contribution to the overall balance of happiness—and that is problematic. By contrast, a strongly held moral principle is that people are valuable in themselves.

This idea of individuals being valuable in themselves is the central idea in deontological or Kantian theory. However, before we take up this alternative theory, we should note that utilitarianism goes a long way in providing a reasoned, systematic, and comprehensive account of morality. It is not a theory to be dismissed lightly. The idea of basing moral decisions on something that is intrinsically valuable is compelling; in some sense, all human beings seem to seek something like happiness or well-being; and consequences are an important element in moral reasoning and in moral practices.

Kantian Theory

In utilitarianism, what makes an action right or wrong is outside the action: it is the consequences of the action. What the person is thinking at the time of the action is irrelevant—that is, little importance is given to what a person intends or seeks to achieve by an action. By contrast, Immanuel Kant (1724–1804) put forward an ethical theory that is focused on the internal character of actions. Because his theory emphasizes acting from duty, it is characterized as a deontological theory. (The word *deontology* derives from the Greek words *deon*, duty, and *logos*, science.) According to Kant, actions that are done from a sense of duty are morally worthy, and those done for self-interest or simply out of habit are not. Thus, if I tell the truth because I fear getting caught in a lie or because I believe that people will think better of me if I do, then my act is not morally valuable. By contrast, if I am in a difficult situation in which telling the truth may get me into trouble and I tell the truth because I recognize that I have a duty not to treat the person asking the question as a means to my well-being, then my action is morally worthy.

Kant gives a thorough account of what it means to act from duty, but before explaining that, it will be helpful to return to utilitarianism. Utilitarianism goes wrong, according to deontologists, when it focuses on happiness as the ultimate good for human beings. Happiness cannot be the highest good for human beings because it is not a unique or special capacity of humans. Humans have the capacity to reason. We are rational beings. This is not to say that humans always do the most rational thing. Rather, the point is that human beings think about how they will act and then choose how they will behave. Many animals and all inanimate objects are entirely controlled by the laws of nature. Plants turn toward the sun because of photosynthesis. Objects fall because of gravity. Water boils when it reaches 212° F (100° C). In contrast, humans have ideas and can make rules and follow them. I can decide to (or not to) exercise every day, lie when asked a question, refuse charity to a beggar, keep my promises, or pay my debts on time. According to Kant, this is the difference between acting according to law and acting according to the conception of law. Humans are unique insofar as we can act according to the conception of law—that is, according to rules and laws that we recognize.

Importantly, if humans lacked the capacity to reason about their behavior before acting, morality would not be possible. Morality is a set of rules and values that humans adopt to direct their behavior. We can act on our beliefs about how we should behave. Morality can't apply to entities that act by instinct or whose behavior is entirely determined by the laws of nature.

Of course, questions arise about how we should think about infants, those with mental disorders, and animals with varying degrees of intelligence, but the important point is that for Kant, the capacity to give laws to one's self is the foundational principle of morality. Moral behavior both derives from the rational capacity of humans and directs it. Morality would not be possible were it not for the fact that humans have the capacity to choose how to act. If our behavior was entirely controlled (determined) by nature, then we would not be capable of following moral edicts.

For deontologists, the fact that human beings have this capacity means that we should recognize this capacity in every human being—that is, we should treat human beings as beings with the capacity to control their own behavior. This idea is captured in what Kant calls the *categorical imperative*. There are several versions of the imperative, but the first expresses the most fundamental idea: We should never treat another human being merely as means; we should always treat others as ends in themselves. When you treat someone merely as a means to your own ends, you deny their humanity; you deny their capacity to reason for themselves and make choices for themselves. This is precisely what is wrong with slavery and other utilitarian practices that treat some human beings as a means to the happiness of other human beings. Such practices treat human beings as things, not recognizing their humanity.

It is important to note that the categorical imperative does not say that you should never treat another person as a means. It specifies only that you never treat another *merely* as a means. You can use another person as a means to one of your ends as long as you do so in a way that recognizes the person's humanity. This means many things, but most important, it means respecting each person's autonomy. To illustrate what this means, consider two of the most significant ways in which you can violate the categorical imperative. One is by lying, and the other is by coercing. Suppose that as an engineer you agree to do work for a client. You might, in a sense, be using the client as a means to your end of making money. However, you wouldn't be treating the client merely as a means as long as you didn't lie to or coerce the client. If you lied to the client about, say, your credentials or the safety of the work that you did, then you would not be respecting the client as a being with her own ends. The client has the capacity to decide whether to hire you or to accept your work, and by lying to her, you have manipulated her into serving your ends and have undermined her capacity to serve her ends. Coercion has a similar problem. Engineers rarely coerce their clients, but extortion comes close. Extortion is an example of pressuring a client to do something that he wouldn't choose to do unless threatened. Say, you threatened to harm a client or one of his loved ones unless he gave you a contract to do work. Again, you would be using the client for your ends and not treating the client as a being with his own ends, able to make decisions for himself.

Another way that Kant proposed that we think about the categorical imperative involves universalizability. Kant argued that, implicitly, actions have maxims. A maxim is the principle underlying an action. For example, suppose you promise a contractor to pay her on completion of work on a building contract and then you decide not to pay her right away (say, you discover you had less money in your bank account and would fall below the minimum balance for earning interest if you paid her right away). Were you to try to universalize your action, it would seem that the underlying principle is that it is okay to break a promise when it is inconvenient or somewhat costly to keep it. According to Kant, you should only do those actions the maxim of which you could universalize. You would be unlikely to want to universalize this principle because that would mean that it was okay for everyone to renege on their promises when it was inconvenient to keep them. Among other things, this would mean that it is okay for those who promised *you* something to renege on their promises. However, Kant's point is not that it is against your interest to universalize this maxim. Rather, universalizing this maxim would, according to Kant, undermine the very meaning of promising. If breaking promises were routine and commonplace, uttering the words "I promise" would be meaningless. And if you can't universalize the maxim of an action, then the action is wrong. Using the categorical imperative in this way, Kantian theory asserts that some actions are always wrong, including promise breaking, lying, and killing—because the maxims of such actions cannot be universalized.

Like utilitarianism, Kantian theory has its critics. As might be anticipated, one line of criticism is to insist that the consequences of actions are important and should not be ignored. Rather than go further into the debate between utilitarians and Kantian theorists, we will turn to another theory that faults both utilitarianism and Kantian ethics for their blindness to the role of virtues in understanding morality. Before we do that, however, a note about a contemporary technological manifestation of the debate between utilitarian and Kantian theory is worthy of mention.

Utilitarianism and Kantian Theory in Autonomous Cars

The difference between utilitarianism and Kantian theory has recently come into play in a controversial engineering problem. In the public discussion and debate about autonomous cars, some attention has been given to how the cars might be programmed to respond in life-threatening situations. Suppose that two autonomous cars are about to crash. The crash is unavoidable, but there is time for some maneuvering in how the cars collide. Sensors make it possible to calculate how many people are in each car and whether any are children. The software of such cars might be designed, following a utilitarian principle, to communicate with the other car or cars involved and have each swerve to minimize the total harm done to all those involved in the accident. On the other hand, some have argued that the software in each car should be designed to minimize the harm done to passengers in that car. In other words, in deciding how to swerve, the software would only consider the car's passengers and not give priority to any of the persons involved. This seems Kantian in that it treats the passengers in each car as valuable in their own right—that is, as ends in themselves. In support of the Kantian approach, some argue that consumers of autonomous cars wouldn't buy cars that would sacrifice them for overall net good.

This issue may well have to be resolved as autonomous cars become more sophisticated. This will be intriguing to follow since neither ethical theory is simple and, as suggested above, each has ways of dealing with criticisms. Chapter 7 is focused on safety and autonomous cars as an emerging technology.

Virtue Ethics

Both utilitarianism and Kantian ethics are focused on actions: utilitarianism on the consequences resulting from an action, and Kantian ethics on the internal form and motivation of the action. By contrast, virtue ethics focuses on traits of character— the virtues.

According to virtue theory, in thinking about ethics, we should focus on the characteristics that make for a good person. Although there are many virtue traditions, virtue ethics in the West origi- nates with Aristotle (384–322 BCE) in his treatise *Nicomachean Ethics*. Aristotle is a teleological thinker. He argues that in order to find out what is good, we should start by asking about the end or purpose or function (the *telos*) of the entity or activity at issue, which may involve a distinctive function (*ergon*). We can then identify the characteristics that make the entity or activity good for its end. What makes something good is that it has character- istics that are suited to achieve its purpose. For example, if you want to know what constitutes a good automobile, you start by asking what the purpose of a car is. We generally think that a car is supposed to transport a person conveniently and efficiently from one place or another. We also acknowledge that some people select cars in order to make a statement about their personal identity. So, a good car will be one that is easy to drive, safe, and efficient and makes an appropriate personal statement. These are the excellences of automobiles; automobiles that have these characteristics are best for what cars are supposed to do.

When it comes to ethics, virtue ethicists propose that we use the same reasoning. Ethics has to do with people, so our attention should be focused on what it means to be a good person. To understand this, we should ask: What is the end or essential function of being a human being? And what are the traits that make for living a good life—that is, for flourishing as the essentially social animals that we are? For virtue ethicists, the proper goal of human life, as specified by Aristotle, is *eudaimonia*. This word is typically translated as happiness or well-being or human flourishing. Flourishing is not merely subjective happiness or good feeling; it is an objective and social condition of being and acting well with others. This end is realized through *arete* (the virtues). The virtues are those traits of character that are essential to human well-being, but they are not merely instrumental to human flourishing: they constitute it. A person who reliably exhibits virtues will have a good life.

The virtues are dispositions to behave in a particular way, as appropriate for a particular situation or context. For example, honesty is a virtue, and a person who is honest is one who has the disposition to tell the truth in ways that are appropriate to each situation. Virtues have an enduring quality—one who is honest doesn't just tell the truth on occasion or only when it is convenient. An honest person is one you can count on to tell the truth when the situation calls for it, and to do so in the right way. For example, a person who spreads gossip or inappropriately betrays others' secrets would not be virtuous, even if they were technically telling the truth. An honest person is consistently honest because honesty is part of the person's makeup. The virtues can be thought of as the excellences of a human being. They are the qualities that constitute excellence in a human being. A virtuous person will flourish as a human being. A life in which one behaves with honesty, courage, and generosity, for example, will be a good life, a life constituted by *eudaimonia*.

Virtues are not, of course, inborn; they are developed through practice and habit, and later refined with the aid of one's practical reasoning (*phronesis*). Virtues must be cultivated. They can be developed to some extent as one is growing up, but we all have to work on ourselves to become virtuous. The challenge of being a good person is that of cultivating in yourself the dispositions essential for virtuous action. As one develops, strengthens, and refines these dispositions, one achieves *eudaimonia*.

In recent years, practical ethicists have used virtue theory to think about virtues not just in general but as they may apply to acting in particular roles, such as being a manager or a parent or an engineer. The idea is that when one finds one's self in a situation requiring moral judgment or decision-making, rather than simply looking for a rule or principle or decision procedure to follow, one should consider the virtues associated with being a good X, where X is a particular role.

Although this way of thinking may not always dictate the specific course of action to be taken in a situation, the virtuous person's practical reasoning will help direct the person to an appropriate form of behavior. For example, if an engineer is in a situation in which a client has asked her to do something that she thinks is unsafe, focusing on the virtues of engineering would lead the engineer to frame and understand the situation in a particular way. It would involve thinking about the ends of engineering and the virtues that would constitute a good engineer. It seems that a virtuous engineer is one who cares about the safety and welfare of society. It also seems that courage is an important virtue for engineers insofar as they have to tell others, such as clients, what they don't want to hear. Clients may not understand technical matters or may give lower priority to safety than they should. So, courage is important for engineers to cultivate in dealing with clients.

Several engineering ethicists have used virtue ethics to think through the virtues specific to engineers. Charles Harris, for example, argues for virtue ethics as a counterbalance to the limitations of codes of ethics.[1] He distinguishes two kinds of virtues in engineering, the technical and the nontechnical. In discussing the technical excellences of being an engineer, he mentions the obvious importance of mastery of mathematics and physics, engineering science, and design as an important virtue, but he also mentions sensitivity to risk and sensitivity to tight coupling or complex interactions as technical virtues. For the nontechnical virtues, he discusses the importance of techno-social sensitivity, respect for nature, and commitment to public good. According to Harris, these are the dispositions that engineers should cultivate to become good engineers.

Virtue ethics is a third framework for thinking through ethical issues. Ethicists often debate which is the better of the three theories just discussed, but in practice very few people adopt one of these theories whole cloth as the basis for living their lives. The three approaches are better understood as frameworks for ethical analysis. They are different ways of thinking and provide concepts and language that must be interpreted and extended to think through real-world situations.

PRACTICAL ETHICS: ETHICS AS DESIGN

As the autonomous car dilemma suggests, real-world ethical dilemmas are generally more complicated and require much more nuanced and iterative thinking than ethical theories alone can provide. This is both because the details of a situation matter a good deal and because real-world situations take place in time and require multiple and sequential actions. To use a simple example, if you discover something to suggest that your employer is not meeting safety standards, you have to decide whether to report this and to whom. Once you have made that initial decision and acted—say you decide to report it to your supervisor—the problem may not be over. Suppose your supervisor tells you that he already knows and doesn't think it warrants further action. You then must decide whether to do something else or nothing. Even if you do nothing, the situation may change if you find additional evidence that justifies your concern or if others begin to see the problem. The point is that you may have to make decisions and choose actions at many points in an ongoing process.

To understand practical ethical decision-making, many ethicists recognize that although theories are useful, they aren't the be all and end all. Ethical decision-making often requires a form of creativity. Instead of deducing the correct answer from a theory, one must synthesize many elements, make trade-offs, and design a solution.

Caroline Whitbeck developed this conception of practical ethical decision-making, especially in engineering, by making an analogy between practical ethical problems and design problems.[2] When it comes to design problems, there are solutions that are blatantly wrong but there generally is not a single right answer. That is, there may be several different designs that meet the design criteria and solve the problem. Each solution solves the problem in a different way, and all may be adequate.

Whitbeck asks us to imagine multiple teams of engineers who are tasked with designing a car seat for infants. Imagine that each team is given the same set of specifications for the design—that the car seat should meet regulatory requirements, not weigh more than a certain amount, fit into a certain percentage of automobiles on the road, cost no more than X to manufacture, and so on. Because of the design criteria, certain designs simply will not work— that is, they won't meet the specifications. On the other hand, each design team may come up with a unique design that fits the criteria. The differences in the designs result from different trade-offs. The teams balance cost, weight, size, safety, and ease of manufacturing differently. One team opts for a lighter seat even though it is more expensive (though still within the specs for cost); another team opts for extra safety (beyond what is required) though this will make the seat heavier than it might otherwise be (yet still within the specifications for weight); a third team might opt for ease of use but more difficulty manufacturing; and so on. The point is that although there are some designs that won't work at all, there is no single right or perhaps even best design choice. Each design will have advantages and disadvantages over other designs.

So it is, Whitbeck argues, with ethical decision-making in the real world. In any situation, there are decisions that are clearly wrong. For example, an engineer shouldn't lie about the safety of a product or take bribes from suppliers. On the other hand, if an engineer discovers an unsafe situation, there is usually more than one possible first step that might be taken to handle the situation. Doing nothing wouldn't be an option, but the engineer might go to her supervisor, or she might go to an ombudsperson or to a senior trusted engineer outside the organization to find out if the senior engineer agrees with the engineer's technical judgment. Depending on the outcome of one of these initial actions, the engineer would have to decide on the next step. Obviously, choices at each step in the process will have advantages and disadvantages that must be balanced, and how they are balanced will depend on such factors as how serious or dangerous the situation is, how time sensitive the problem is, the probability that each course of action would be successful, and so on.

Thinking of practical ethics as design doesn't make ethical decision-making easy. It still requires significant thought and analysis and attention to the details of a situation. Nevertheless, viewing ethical problems as design problems is closer to the real-world experiences of engineers as they try to balance many factors and consider the uncertainties that go with each step that they take. In this way, viewing practical ethical problems on the model of design problems acknowledges that ethical choices are not simple matters of deducing the right answer from a theory. Engineering ethics decision-making involves judgment; solutions and courses of action are often synthetic and creative.

CONCLUSION

So, how should engineers think about ethics? In this chapter we have suggested several ways (not mutually exclusive) that engineers can and should understand the ethical dimensions of their work. First, engineers should think about engineering ethics as professional ethics but not in a narrow sense. Engineering ethics is not just a matter of following professional codes of conduct. At the core, engineering ethics has to do with how engineers behave in the many social relationships involved in engineering and how they consider the social effects and implications of the work that they do. Second, in dealing with these ethical dimensions, engineers can and should draw on ethical theories. Ethical theories provide a systematic way of thinking about ethics that can be used to understand the situations in which engineers find themselves. They provide frameworks, concepts, and vocabulary that allow engineers to see their circumstances and dilemmas in ways that they might not have comprehended without theoretical analysis. Third, engineers sometimes find themselves in situations in which some options are clearly wrong but the right thing to do is either not obvious or not possible without compromises or trade-offs. So, it is important for engineers to understand that practical ethical decision-making is like design. To figure out what to do in some situations, engineers must design courses of action that involve taking a wide variety of factors into account and acting one step at a time. In the next chapters, we will explore professional values and relationships and how engineers can best balance their many responsibilities.

SUGGESTED FURTHER READING

Johnson, Deborah G., and Jameson M. Wetmore. "STS and Ethics: Implications for Engineering Ethics." Pages 567–82 in *Handbook of Science and Technology Studies*, 3rd ed. Edited by Edward J. Hackett et al. Cambridge, MA: MIT Press, 2008.

Rachels, James, and Stuart Rachels. *The Elements of Moral Philosophy*, 8th ed. New York: McGraw-Hill, 2015.

Whitbeck, Caroline. "Ethics as Design: Doing Justice to Moral Problems." *Hastings Center Report* 26, no. 3 (1996): 9–16.

PART TWO

Employment Relationships

4 SHOULD ENGINEERS SEE THEMSELVES AS GUNS FOR HIRE?

WORKING as an engineer involves entering into a variety of social relationships—with managers and supervisors, clients, contractors, inspectors, suppliers, and other stakeholders who are affected by one's work. In the last category might be residents of an area affected by a structure that an engineer is helping to build or consumers/users of products that an engineer has helped to develop. The success of an engineer's career depends in critical ways on effectively managing these relationships.

Two of the most important relationships are those with employers and those with clients. Both are employment relationships in the sense that it is by working for others that engineers earn a living. Employment relationships can take many and complicated forms, but generally engineers either work for themselves as consultants to clients or they work as employees of organizations such as corporations, government agencies, and nonprofits. In this chapter, we will explore some of the ethical issues that arise in engineer-client relationships, and in chapter 5, we will turn to engineer-employer relationships.

Engineers work with clients in variety of arrangements, and the ethical issues that arise vary somewhat with the arrangement. For example, self-employed engineers may have multiple clients, and this creates the potential for conflicting responsibilities when the interests of one client clash with the interests of another client. Imagine, for example, that an engineer is asked by a new client to implement a manufacturing process the client has just developed. The engineer knows from working with other clients in the same industry—competitors of the new client—that the new client's system is out of date and inferior to the other client's. Were the engineer to reveal this to the new client, she might violate the confidentiality of her other clients and undermine their competitive advantage. In another form of the employment arrangement, engineers work as employees of organizations that have clients. Here engineers are in engineer-client relationships, but the clients are clients of the engineer only insofar as they are clients of the company for which the engineer works. Here situations can arise in which there is a conflict between the interests of the client and the interests of the engineer's employer. For example, an engineer might see a way to solve a client's problem that would mean very little future business for the engineer's employer-company. Hence, the engineer faces the dilemma: Should I tell the client what will serve his or her interests or keep quiet to protect the interests of my employer?

As indicated in chapter 2, most engineering codes of ethics specify that engineers should be faithful to their clients. Clients hire engineers and engineering firms to make use of their engineering expertise, and they expect engineers to use that expertise to serve their (the clients') interests. However, as already suggested, a variety of situations can make it difficult for an engineer to balance fidelity to the client with other responsibilities.

To explore the complexities of engineer-client relationships and the nature and limits of an engineer's obligations to clients, we can consider a view of engineers that is antithetical to the idea of engineers as professionals. Individuals who provide their services to anyone who is willing to pay and don't ask or care about the project and how it will affect society are the equivalent of *guns for hire*. Guns for hire do whatever their clients want. Guns for hire leverage whatever talent or expertise they have for their personal gain. Imagine a young engineer seeing her- or himself as having worked very hard to acquire their engineering expertise and now wanting to sell themselves to the highest bidder. They may even argue it is not their place to decide what projects get pursued—that is the prerogative of the client, that is, the one who pays. On this view engineers are the means, and it is up to their clients to determine what ends are sought, what goals are achieved. Guns for hire believe that expert knowledge is not just one of the distinguishing features of engineers: it is the only distinctive feature. So, viewing engineers as guns for hire would mean viewing them as individuals who are free to leverage their knowledge in whatever way that suits them.

By contrast and as explained in chapter 2, professionals see themselves as having special expertise as well as a commitment to protecting the public. As professionals, engineers see themselves not just as individuals but as members of a profession with responsibilities to the public and to maintain the integrity and the trustworthiness of the profession. According to engineering codes of ethics, engineers are to hold paramount the health, safety, and welfare of society and to uphold certain standards of conduct. This is the antithesis of a gun for hire.

In this chapter, we take up the question whether engineers should be viewed, and should act, as guns for hire. As we explore some of the ethical dilemmas that arise in engineer-client relationships, the chapter will argue against this view. Engineers should be viewed as, and should view themselves as, professionals who must balance their personal interests and the interests of their clients with the good of the public and their profession.

PUBLIC SAFETY, CONFIDENTIALITY, AND HONESTY

To begin, consider a case that saliently illustrates a problem with the guns-for-hire view of engineers. This case is from the NSPE compendium of cases reviewed by its Board of Ethical Review:

Engineer A, a fire protection engineer, is retained by Client to provide a confidential report in connection with the possible renovation of an old apartment building owned by Client. Engineer A conducts an audibility test of the fire alarm inside occupied residential units. The audibility test showed the alarm could not

be heard within all of the residential units, which is in violation of
the local fire code. The problem with the alarm may have existed
since the time the building was constructed or when the fire alarm
system was replaced in 1978.

Engineer A advises the Client regarding the results of the
audibility tests and the code violation. In a follow-up telephone
conversation with Client, Engineer A is told that the financing for
the renovation has fallen through and that the renovation project
will be delayed, which means that the problems with the fire alarm
system will not be addressed immediately but in the future when
funding is available. Engineer A is paid for his services.[1]

The question posed to the BER is this: What are Engineer A's
obligations under the circumstances?

In its commentary, the BER acknowledges that the engineer
has an obligation to be faithful to the client, and this entails not
disclosing confidential information concerning the client's business affairs. At the same time, the BER recognizes that this obligation is in tension with Engineer A's obligation to hold public
safety paramount. The BER points out that situations of the kind
described in this case will vary and that an engineer must exercise
judgment in determining how much risk is posed by a situation
like that of the fire code violation. The board takes the position
that the engineer should determine whether the alarm poses an
"imminent and ongoing risk" to the health, safety, and welfare of
the building occupants. If it does, the engineer should go back to
the client and advise the client that appropriate steps be taken to
remedy the situation. If the client refuses to do this, the engineer
should report the code violation to enforcement officials.

In rendering this recommendation, the BER does not exactly
provide a definitive or categorical solution to the problem; it says
that what the engineer should do depends on judgment, and it
recommends a course of action that may or may not work and
then a second course of action depending on the results of the
first. In these respects, the BER recommendation illustrates two
points made in chapter 3 about the nature of ethical decision-
making. Remember that ethical decision-making was described
on the model of solving a design problem; there may be solutions

that don't work—the equivalent of wrong actions—but there may not be one perfect solution—one right way to act. In this case, it seems that it would be wrong for the engineer to do nothing and simply walk away from the job. On the other hand, the right thing to do depends on a reconsideration of the degree of risk to the building occupants and, if the risk is serious enough, then a clear course of action—go back to the client. The next steps are less clear because they depend on the client's response. So, we see that ethical decision-making is not always a simple matter of figuring out *the* one right course of action. It is often an iterative process involving a series of decisions, each dependent on the outcome of the previous, working toward a positive outcome.

Notice that the BER is not thinking of the engineer as a gun for hire. A gun for hire would be happy to receive payment from the client and forget about the inadequacy of the fire alarm. Perhaps such an engineer would say to him- or herself that it was the client's call. One of the reasons the guns-for-hire view is problematic is that it doesn't recognize the relevance of engineering expertise to ends as well as means. Here the client is deciding what is most important; to make an informed decision about whether to do something or nothing, the client needs to understand the nature and degree of the risk. The engineer understands the nature and level of the risk better than the client, so the client could benefit from the engineer's expertise. If the engineer just walks away, the client doesn't get the full benefit of the engineer's expertise.

Although the BER does not adopt the guns-for-hire view of engineering, notice that it doesn't suggest that Engineer A should go *directly* to the authorities to protect the building occupants. In this respect, the board acknowledges an engineer's responsibility to clients. In this case, the engineer's responsibility to his client is recognized in two ways. First, the BER allows that if the risk or threat is not "imminent and ongoing," then the engineer might do as the client wishes. Second, it suggests that the engineer go back to the client and try to convince the client that something should be done. The implication here is that the engineer should not simply take the matter into his own hands and go to the authorities. He owes something to his client and that is why he should go back and try to convince the client to address the safety issue. So, the BER gives weight to the obligation to be faithful to one's client, though it doesn't think that this obligation trumps the engineer's duty to protect the public.

Confidentiality

Engineer A's obligation to be faithful to his client is an obligation to keep information about the client confidential. Reporting the unsafe alarm system to authorities would violate confidentiality. The suggestion in the BER analysis is that confidentiality can be violated in certain circumstances, but this should not be taken as an indication of unimportance. Confidentiality is an important aspect of engineer-client relationships for many reasons, but especially because it involves trust. Clients expect engineers to keep certain kinds of information confidential because doing so protects their interests. When engineers violate confidentiality, they undermine the trust that is essential to engineer-client relationships. Often engineers are asked to sign nondisclosure agreements that commit them to not revealing proprietary information or information that gives a client a competitive edge. However, even when there are no signed agreements, engineers are expected to keep certain kinds of information confidential, and when they fail to do so, clients are less likely to put their trust in those engineers.

To explore confidentiality further, consider another NSPE case. Unlike the first case involving public safety, in this case, the BER concluded that it would be wrong for the engineer to break confidentiality:

> Engineer A works for a government agency involved in the design and construction of facilities. During Engineer A's tenure with the government agency, Engineer A receives access to confidential and proprietary design information provided by companies seeking approval from the government agency for their facility designs. Company X is among the companies submitting confidential and proprietary design information. Engineer A ends her employment with the government agency and accepts an engineering position with Company Y, a competitor of Company X.[2]

The question posed to the BER is: What are Engineer A's ethical obligations under these circumstances? The information that Engineer A acquired while working for the government agency was confidential, and yet Company Y has an interest in that information and may have hired Engineer A precisely because she has this information.

Importantly, the analysis of the BER focuses not just on the confidentiality of the information that Engineer A has but also on whether she should even accept a job with Company Y. Some might argue that Engineer A should not be allowed to work for Company Y because, while working there, she may be tempted to reveal or may inadvertently reveal confidential information about Company X. However, the BER does not take this position and instead concludes that it would "not be unethical" for Engineer A to take a job with Company Y. Nevertheless, the board insists that in her new position, Engineer A must not disclose any confidential and proprietary design information she learned during her employment with the government agency.

Advocates of the guns-for-hire view of engineering might argue that whatever information Engineer A acquires during her employment in a government agency should be considered hers. They might argue that she should be able to leverage whatever information she acquired. That is, they might say, how engineers build their careers; the more they know, the more attractive they are to future employers.

The counter to this is easy to see. Were engineers to routinely reveal confidential information about their clients, there would be long-term negative consequences for individual engineers and for the profession as a whole. Were Engineer A to flagrantly ignore confidentiality in this situation and others, she would likely acquire a reputation as someone who can't be trusted to keep information confidential; clients would be reluctant to hire someone with such a reputation. Similarly, were most or even many engineers to ignore confidentiality, then the profession of engineering as a whole would be harmed. Those who need engineering expertise would be reluctant to give engineers information they might need to do their jobs. Clients would look elsewhere for people with the requisite expertise; they might single out certain trustworthy engineers, but they would not assume that engineers qua engineers will serve their interests.

As should be obvious, this is a rule utilitarian argument. Even though an engineer might personally benefit from a single situation in which he or she reveals confidential information and even though in this particular situation overall good consequences might result, the long-term consequences of such a practice are negative. Good consequences result when engineers make it part of their practice to keep client information confidential.

The practice of keeping information confidential can also be thought through in terms of virtue theory and the categorical imperative. The habit of keeping one's commitments to clients is a virtue that should be cultivated in engineers; it constitutes engineering as a worthy endeavor. The categorical imperative also seems to apply here. Were Engineer A to take the job and reveal information about Company X to Company Y (her new employer), she would in effect be treating Company X and the government agency for which she worked merely as means to her own self-interest. That is what guns for hire do.

Of course, we have to be careful here for although the practice of keeping client information confidential is a good thing, one that engineers should honor and one that is vital to engineer-client relationships, we saw in the case of the fire protection engineer that obligations of confidentiality are not absolute and unqualified. In certain circumstances and especially when public safety is at stake, engineers are justified in breaking confidentiality. One of the benefits of engineering codes of ethics is that they make clear to clients that engineers hold the safety and welfare of society paramount. Because of professional codes of ethics, clients are on notice that engineers may break confidentiality when the public good is at stake.

Honesty

Honesty might be seen as the opposite of confidentiality in the sense that it involves transmitting information rather than keeping it private. However, honesty is like confidentiality in being important to the reputation of engineers and in often coming into tension with engineers' responsibility to hold the health, safety, and welfare of society paramount. Consider this fictional case:

> Residents in the neighborhood of a manufacturing plant have been complaining about what seems to be environmental damage to a river near the plant. A local government agency requests that the company do an assessment of the environmental effects of the plant's waste disposal activities. The company hires Engineer X to do the assessment and asks for a preliminary report. Engineer X does an initial assessment and finds that there is more eco-logical damage than the company projected when the plant was initially built. The preliminary results of Engineer X's assessment suggest that wildlife in and around the river are being seriously and negatively affected. In doing the assessment, Engineer X thinks of several changes the plant could make to reduce the damage, though these changes would be costly. Engineer X writes a preliminary report telling the company what he found, including the recommendations for a remedy. The company responds by asking Engineer X to write the final report in a way that will downplay the damage, making it appear less serious than it is. The company also requests that the Engineer X leave off any mention of the possible remedies. Engineer X worries that if he skews the report as the client suggests, he would be lying. Engi-neer X worries further that if he refuses to do what his client suggests, his reputation as a consultant might be damaged. On the other hand, if he does what the company wants and the truth later comes out, his reputation could be damaged.

Were we to adopt the guns-for-hire view of engineering, we might simply say that the engineer should do what the client wants. Notice that although the case looks similar to the fire alarm case in the sense that it pits a client's interests against the public good (in this case, environmental harm), in this case the engineer's integrity as an engineer is at issue. The client seems to be asking Engineer X to lie or at least to distort the facts.

Honesty is important for engineer-client trust because clients count on engineers to be honest and accurate about the issues they have been hired to address. In this case, the engineer has been honest with the client in his report, but the client wants the engineer to be less than honest in a report that will go to local government authorities. So, it is not honesty with the client but public trust that is at issue. Pubic trust in engineering is undermined if engineers distort the truth when their clients ask them to or when it is convenient or in the engineer's interest. Moreover, the reputation of individual engineers can be undermined if they are known to distort the truth to serve their clients' interests.

Here again we have a situation where one of the engineer's options is clearly wrong but there are several alternative possible courses of action. To outright lie or distort the facts in writing the report would be wrong, but the right or best course of action is somewhat harder to determine and iterative. The engineer may have to take an initial action and wait for the reaction before deciding on the next step. Leaving out of the report the information about possible solutions to the problem would seem not to be wrong because neither the company nor the government agency requested this information. Still, Engineer X is in a situation often described as being between a rock and a hard place. His reputation could be damaged if he refuses to change the report and the company tells other potential clients that they were not happy with Engineer X's work. On the other hand, Engineer X's reputation will be harmed if he writes an inaccurate report and the truth comes out later. As in the fire alarm case discussed before, it seems best if the engineer goes back to his client and explains the seriousness of the situation and figures out a way to keep the report honest. If he does this, his next action depends on how the client reacts. If the company accepts the report without the recommendations as to what to do to remediate the situation, then he may not do anything more. If, however, the company still wants Engineer X to change the report further, he must figure out what more to do. Engineer X might consider whether there are other individuals in the company—besides those with whom he has been dealing—to whom he can go and explain the situation. Is there an ombudsman for the company? Is there an experienced engineer inside or outside the company with whom he could talk over the situation? Does a local engineering professional organization offer advice? Just what he should do depends on what more he learns and further details.

The challenge faced by Engineer X in this case suggests that the guns-for-hire position is not just inaccurate but may be incoherent. On the one hand, if Engineer X were purely out for himself, it seems that he should write whatever his client wants him to write. On the other hand, it is not in Engineer X's long-term best interest to acquire a reputation for dishonesty. That would likely lead to his having trouble finding work. In this respect, the guns-for-hire view of engineering seems self-defeating; if engineers do just what their clients wish, they will likely develop reputations as dishonest and unreliable engineers, engineers without integrity, and that is not in their long-term interests. Yes, there may be clients who want sleazy engineers and there are probably some engineers who think only about their short-term interests, but this leads to corruption and is dangerous for clients and engineers as well as others. We will take up the topic of corruption later in this chapter.

CONFLICTS OF INTEREST

Conflicts of interest are another category of situation that can be problematic for engineer-client relationships. A conflict of interest situation is one in which: (1) an individual acts in a professional role such as engineer; (2) individuals in that role are expected to exercise judgment or make decisions of a particular kind; *and* (3) the individual has an interest that could interfere with the expected kind of judgment. When engineers act in conflicts of interest situations, the trustworthiness of the individual engineer and the profession is undermined.

Of course, engineering is not the only profession in which conflicts of interest arise, and it may be helpful to begin with some professions in which the harmfulness of conflicts of interest is blatantly obvious. Consider the case of judges; judges are expected to render judgments that are fair, and this means that they should be impartial. Suppose a judge is assigned to preside over a case involving a dispute between two companies and the judge has a significant financial investment in one of the companies. In that situation, the judge has a conflict of interest and would be expected to recuse herself from the case. Similarly, an umpire would be considered to have a conflict of interest were he to be assigned to officiate at a game in which his child or spouse was a player. The judge is expected to judge on the merits of the case, and the fact that she has a financial investment in one of the companies raises the question whether she can judge in a disinterested manner. Likewise, the umpire is expected to make determinations without bias toward one team or team member and another, and the fact that a family member is on one of the teams raises the question whether he can make calls that are fair. Both the judge and the umpire have personal interests that could interfere with their rendering the kind of judgment expected.

When it comes to engineers, *objectivity* is the term often used to describe the kind of judgment that is expected; engineers are expected to be objective in their evaluations and recommendations. Engineers are expected to make decisions that are objective in the sense that they are based on the engineer's technical expertise and not the engineer's personal interests. The most common types of interests that raise conflict of interest questions are financial, familial, and romantic. These are personal interests that can get in the way of objectivity. However, it is not just personal interests that threaten objectivity; conflicts of interest can also arise when engineers have multiple clients and the interests of one client compete with the interests of another client.

One point of clarification is important here. Conflict of interest situations are different from situations in which one must balance *conflicting interests*. Like all persons, engineers have lots and lots of interests—in their personal finances and their families, as well as in their friends, churches, sports teams, political parties, ethnic groups, and so on. Most of these interests are not relevant to engineering judgment and decision-making, so they are not relevant to conflicts of interest. Conflicting interests usually involve balancing time and effort. For example, an engineer may have personal responsibilities that involve taking care of an elderly parent or coaching a children's baseball team, and these responsibilities may be in tension with the time and effort necessary to perform work. These interests do not, however, conflict with the character of the judgment that engineers provide. Thus, situations of this kind are not referred to as conflicts of interest. The term *conflict of interest* refers specifically to situations in which an interest could interfere with, or could be perceived as interfering with, the objectivity of engineering judgment.

To see the problem with conflicts of interest, consider this fictional case:

> An engineer is hired by a university to help determine the best software system for providing websites to support courses taught by faculty from a wide array of disciplines. The engineer is tasked with evaluating software packages that are available for this purpose and identifying the one that is best suited for the university's needs. The engineer does the evaluation carefully, comes up with a recommendation, presents it in a written report, and collects his consulting fee. He recommends that the university buy the software system offered by Company ABC, and he gives extensive reasons for this recommendation.
>
> The engineer does not mention that he used to be an employee of a company that developed software of this kind. Of course, it was clear from his resume that he had worked for Company ZZZ. What wasn't clear was that ZZZ was bought out some years ago by Company ABC. While working at ZZZ, the engineer made a significant contribution to the design of the system that is now being sold. Moreover, while working at ZZZ, the engineer acquired a substantial number of shares in the company, and in the take-over arrangement, those shares became shares in Company ABC. As a result, the engineer has a significant financial interest in Company ABC.

Although the engineer may believe that his connection to Company ABC did not interfere with his judgment about the best software system for the university, if the university were to discover the relationship, the university would likely wonder whether the engineer's recommendation had been objective. The possibility that the engineer's recommendation had been influenced by the connection to ABC undermines the credibility of the engineer's judgment. Notice here that it is not just the financial interest that is problematic but also that the engineer may be perceived as having a biased perspective on software because he had a hand in its development. At the heart of conflicts of interest is the idea of tainted judgment.

Importantly, conflict of interest refers to a situation that may not be true—that is, a person's interests may not influence her or his judgement. For example, in the case of the judge, if she were to hear the case, she might make efforts to ensure that her personal interest did not interfere with her judgment. Indeed, one problem with allowing professionals to act in conflict of interest situations is that they may bend over backwards so as not to appear interested. Hence, they may favor the party for whom they have no personal interest. Imagine, for example, that an umpire tries to show his impartiality in a game in which his daughter plays. In doing so, he favors the team competing against his daughter's team. This too would be unfair, and it shows that conflicts of interest can harm those with whom the decision maker doesn't have an interest as well as those with whom the decision maker is aligned.

Still, let us press the point further. Suppose that some individuals with a conflict of interest are able, despite their interests, and without bending over backwards in the other direction, to render objective judgments. For example, suppose in the case of the engineer selecting the software package that he is careful not to let his financial interests or his fondness for ABC's design approach to affect his judgment. We can even imagine that the university hires three additional engineers to replicate the engineer's work, and they all agree that ABC's software is the best package for the university. Nevertheless, it would not be good to let engineers (and other professionals) act in situations in which they have a conflict of interest. Conflicts of interest are problematic not just because of the actual conflict but because of the appearance of a conflict of interest. If stakeholders to a decision see that a professional has a conflict of interest, the professional's judgment is called into question. The judgment appears tainted, and this may carry over to other members of the profession. That is, professionals who act with conflicts of interest erode the reputation for trustworthiness of the profession as a whole.

Although the mere appearance of a conflict of interest is problematic, some argue that disclosure solves the problem and promotes trust. Most engineering codes of ethics treat conflicts of interest as something to be avoided, while others specify that engineers need only disclose their conflicts of interest and leave it to the client to decide whether he or she still wants the engineer to do the work. For example, in elaborating the Fundamental Canon on fidelity to employers and clients, the NSPE Code of Ethics addresses the appearance of conflict of interest and recommends disclosure. It specifies that engineers shall "disclose all known or potential conflicts of interest that could influence or appear to influence their judgment or the quality of their services." Similarly, the IEEE code states that members are "to avoid real or perceived conflicts of interest whenever possible, and to disclose them to affected parties when they do exist."

To illustrate the strategy of disclosure, consider another case from the NSPE. This is a case involving a familial relationship:

> Engineer A has been assigned by her engineering firm to work with a local developer on a commercial development project. While performing the work, Engineer A becomes aware that her father, an adjacent landowner, is participating with a community group in an appeal of a zoning board's decision to grant the developer a zoning reclassification of the developer's property in order to permit the building of commercial development. Engineer A's father has the adjoining property and is planning to build a new home.[3]

In the BER's discussion of this case, it concludes that Engineer A "has an ethical obligation to fully disclose this information to both the employer and developer client; it is for those parties to determine whether Engineer A should continue on this assignment." So, the board recognizes that the interest of Engineer A's father in the project is something that could affect her judgment. They presume that her father's interest creates an interest in her, an interest that might affect her judgment. Importantly, however, the BER does not recommend that the engineer automatically remove herself from the situation (as a judge might be expected to do if asked to preside over a case in which the judge's father had a similar connection). Rather, the BER recommends disclosure and leaving it up to Engineer A's employer and client to determine whether they want her to continue in her current role.

As should now be clear, conflicts of interest have everything to do with the trustworthiness of professional judgment. Although we will not go into detail here, conflicts of interest can be understood using the ethical frameworks discussed in chapter 3. The practices of avoiding conflicts of interest, avoiding the appearance of conflicts of interest, and disclosing such conflicts are all justified in a utilitarian framework—that is, good long-term consequences result from such practices. One of those consequences is to bolster trust in engineers and engineering. Although the argument would be different, the harm of conflicts of interest can also be explained using the categorical imperative. In effect, keeping one's potentially interfering interests hidden from one's clients is a way of manipulating them; an engineer's potentially interfering interests are relevant to a client's decision to hire and rely on an engineer. Thus, when engineers act in conflict of interest situations, they use their clients as a means to their own ends. On virtue theory, developing in one's self the habit of avoiding or disclosing conflicts of interest would be seen as a virtue, a virtue that should be cultivated by engineers because it promotes trust. These are all different ways of understanding why it is so important for engineers to manage conflicts of interest situations with care.

CORRUPTION

Acting with or not disclosing a conflict of interest is considered a form of corruption, but corruption is not limited to conflicts of interest. Corruption takes many forms, and engineers can find themselves in situations where they are tempted to engage in corruption, circumstances in which they are pressured by clients or employers to be part of a corrupt practice, or situations in which they observe others engaging in corrupt practices. All forms of corruption, like conflicts of interest, threaten the integrity and trustworthiness of engineers and engineering.

What exactly is corruption? Most definitions refer to dishonest or fraudulent or illegal behavior by those who seek personal gain. Bribery, embezzling funds, using project equipment for personal use, bid rigging, siphoning off funds, or selling goods intended for a project for one's personal gain are all forms of corruption. Usually corruption involves an inducement to do something that is improper. The inducement may take the form of gifts, loans, fees, rewards, or other advantages. These inducements are offered with a view to influencing the judgment or conduct of a person in a position of trust. Transparency International defines corruption simply as "the abuse of entrusted power for private gain."[4]

Corruption occurs in many domains of engineering, but it is especially problematic in the construction industry. This is evident both in the amount of public attention given to corruption cases as well as in the amount of regulation that now exists in that industry. Also, although corruption can and does occur in developed as well as developing countries, it is especially problematic in the developing world. In fact, some observers argue that there is a correlation between corruption and poverty.[5]

To give some salience to the problem of corruption in the construction industry consider reports of a few recent incidents:

According to ConstructionDive, on March 1, 2018, a former Metropolitan Transit Authority (MTA) construction project administrator was ordered to pay a fine of twenty thousand dollars and serve a forty-six-month sentence in prison after pleading guilty to asking for and receiving bribes from MTA contractors.[6]

According to Reuters, in 2018 a former executive of an upstate New York development firm pleaded guilty and agreed to cooperate in a corruption case involving a billion-dollar government project. The executive admitted to wire fraud and conspiracy in a bid-rigging scheme that allowed his company to win a lucrative contract.[7]

According to a 2015 *New York Times* article, Hunter Roberts Construction Group, a major New York construction company, inflated bills by altering the time sheets of project foremen. The additions were small: one hour of overtime here, another two hours there; when foremen went on vacation or took a sick day, the time sheets falsely indicated that the workers had showed up on the job. The article explains that "Hunter Roberts admitted to the conduct and agreed to pay $7 million—$1 million to affected clients and $6 million to the government—in a nonprosecution agreement."[8]

BBC News reported that in 2016 one of the largest cases of corruption in history was settled when Odebrecht, a Brazilian-based construction conglomerate, confessed and agreed to pay $2.6 billion in fines. Seventy-seven Odebrecht executives consented to plea bargains confessing that they had paid bribes in exchange for contracts—not only in Brazil but in a number of other countries around the world.[9]

Although it is difficult to make generalizations about how individual engineers should behave when they are offered bribes, observe others taking bribes, or get caught up in some form of corruption, it is important to emphasize that the most effective strategies for engineers as a group are activities that prevent corrupt practices from developing altogether. For this reason, there are many organizations and activities devoted to ending corruption globally. For example, the World Economic Forum sponsors an organization named Partnering against Corruption Initiative that develops principles and practices to eliminate corruption in project procurement and performance. Companies can formally commit to these principles and practices and implement them in their operations. Another body is the Global Infrastructure Anti-Corruption Centre. This is "an independent, non-profit organization that promotes implementing anti-corruption measures as part of managing companies, agencies and projects."[10] These efforts focus on companies (rather than individuals) and help firms to develop strategies to eliminate corruption within their organizations and within their industries.

Because most forms of corruption are illegal, it may seem unnecessary to discuss this topic. That is, it may go without saying that engineers should not engage in illegal behavior. In fact, many engineering professional codes of ethics do not include statements prohibiting corruption for precisely this reason. As mentioned in chapter 2, some believe that actions prohibited by law need not be additionally prohibited in a professional code of conduct. Yet, another reason for not including prohibitions on corruption is that corruption is essentially about the conduct of business—that is, how business is done. So, it is not specifically about or specific to engineering. Although the latter argument is accurate, it is worth emphasizing that engineering takes place in the context of business, so that business practices significantly affect engineering endeavors. If materials are bought from a company because the company paid a bribe, the materials may not be of the quality necessary for safety; if the bidding process is rigged, an unqualified engineering firm may get the job; if money is siphoned from a project, the cost of the project to citizens or clients may be unfair.

Within engineering, it remains a debatable question whether prohibitions on corruption should be included in engineering codes of conduct. Professional societies have decided the question in different ways.

CONCLUSION

Having explored various dimensions of engineer-client relationships, the answer to the question whether engineers should be viewed, or should view themselves, as guns for hire seems clear. As suggested at the beginning of this chapter, guns for hire are the antithesis of professionals. Engineering is a profession, and members of a profession have responsibilities that go beyond their own short-term self-interest and the interests of their clients. This chapter has argued that acting as professionals and adhering to the norms of the profession serves the long-term self-interests of individual engineers. Professions maintain norms of practice that help to ensure that practice has positive outcomes. An important part of being a professional is managing relationships with clients. As professionals, engineers are obligated to be faithful to their clients, but this does not mean that engineers must do whatever their clients ask. Instead engineers must balance their obligations to clients with their other responsibilities. In certain circumstances engineers are justified in acting against the wishes of their clients, especially when public safety is at stake or when a client asks an engineer to engage in corruption.

SUGGESTED FURTHER READING

"A Question of Ethics." Column published eleven times per year in *Civil Engineering Magazine*. https://www.asce.org/a-question-of-ethics/.

The Royal Academy of Engineering. *Engineering Ethics in Practice: A Guide for Engineers*. London, August 2011. https://www.raeng.org.uk/publications/other/engineering-ethics-in-practice-full.

5 ARE WHISTLEBLOWING ENGINEERS HEROES OR TRAITORS?

WHISTLEBLOWING is one of the most discussed topics in engineering ethics. This is because acts of whistleblowing are a manifestation of a tension at the heart of the role of engineer. As professionals, engineers have strong commitments to protect the health, safety, and welfare of society and to adhere to professional standards, yet at the same time, most engineers work as employees for companies that expect their employees to serve the interests of the company. In a sense, this means that engineers are evaluated by two standards, the standards of the profession and the standards of the business world. Although the interests of employers are not directly or inherently in conflict with protecting public safety and welfare, circumstances can arise in which engineers face a choice between protecting the public or doing what their employers want.

Consider the case of Salvador Castro. According to an article in a 2004 issue of *IEEE Spectrum*, "When Salvador Castro, a medical electronics engineer working at Air-Shields Inc. in Hatboro, Pa., spotted a serious design flaw in one of the company's infant incubators, he didn't hesitate to tell his supervisor. The problem was easy and inexpensive to fix, whereas the possible consequences of not fixing it could kill. Much to his surprise, though, nobody acted on his observation, and when Castro threatened to notify the US Food and Drug Administration (FDA), he was fired."[1]

Something similar happened to Walter Tamosaitis, a nuclear engineer working at the Hanford Waste Treatment Plant in Washington State. In Tamosaitis's case, he expressed his concerns over a long period of time before he was fired. Hanford had been a US nuclear waste site, and in 2000 the Department of Energy contracted with Bechtel Corporation to build a plant that would encase the waste in glass logs through a vitrification process.[2] URS Energy & Construction was a subcontractor to Bechtel on this enormous project. Walter Tamosaitis worked for URS as the research and technology manager.

As early as 2003, Tamosaitis began expressing concerns about design defects that he thought posed serious safety issues. Managers at Bechtel were not happy about him raising these issues, and their dissatisfaction reached a head in July 2010, when they fired him. However, because Tamosaitis was an employee of URS, being fired by Bechtel did not end the matter. Instead, URS reassigned him to a far from ideal post. One reporter wrote that Tamosaitis was "relegated to a basement storage room equipped with cardboard-box and plywood furniture with nothing to do."[3]

Soon thereafter, Tamosaitis reported his concerns to the Defense Nuclear Facilities Safety Board. He also filed a whistle-blower complaint with the Occupational Safety and Health Administration under the Energy Reorganization Act and filed suits against Bechtel, URS, and the Department of Energy. Three years later, in October 2013, Tamosaitis was laid off from URS. During the time between reassignment and being laid off, he continued to speak out and testified before Congress. Studies made of the site identified a number of problems—enough to lead to stoppage of work on certain parts of the plant.[4]

As the Castro and Tomasaitis cases make clear, whistleblowing can be costly to the whistleblower. Both men had their lives thrown into turmoil for many years. Whistleblowing can also be costly to those who are accused by whistleblowers. Organizations targeted by whistleblowers are often hurt by the accusation of wrongdoing. Whether guilty or not, the organization may suffer a tarnished reputation, see its business activity diminished, and have to invest significant resources in investigating the issue and defending itself against the accusations.

In this chapter, the controversy about whistleblowing is formulated into a debate as to whether whistleblowers are heroes or traitors. Of course, the question cannot be answered in its general form. Each case of whistleblowing must be evaluated on its own because the details make a difference: whether a particular whistleblower is a hero or a traitor depends on the particularities of the case. Also, it is important to remember that there is room in between the traitor-hero extremes; for example, one might think that a particular whistleblower wasn't really a hero because he didn't speak out publicly until he was fired or one might conclude that a whistleblower wasn't quite a traitor because although she spoke out in a way that harmed her employer, she did everything she could to address the problem before going public. Nevertheless, formulating a general debate on whistleblowing allows us to explore the ethical dimensions of whistleblowing.

One of the most highly publicized recent cases of whistleblowing is that of Edward Snowden. Although technically not an engineer, Snowden worked for Booz Allen Hamilton doing contract work for the National Security Agency (NSA), a job that an early career engineer with strong computer skills might be hired to do. Snowden, like Castro and Tamosaitis, became aware of serious wrongdoing in the organization in which he worked. However, in Snowden's case, the problem was not a threat to the physical safety of the public: the problem was that the NSA was violating the privacy rights of American citizens. In 2013, Snowden released classified documents that showed that the NSA had been gathering enormous amounts of personal data on Americans who had done nothing wrong. As is now well known, the Department of Justice charged Snowden with violating the Espionage Act and with stealing government property. In order to avoid arrest, Snowden left the country and sought asylum initially in Hong Kong and later in Russia, where he now resides.

Especially because he released classified documents, Snowden is viewed by some as a traitor, and not just a traitor to the organization for whom he worked but to his country. The case might be viewed as atypical because of the way Snowden blew the whistle—that is, by making classified documents publicly available—but the case can also be seen as simply an extreme version of ordinary whistleblowing.

Employers often view whistleblowers as disloyal. Indeed, those who are accused by whistleblowers often attack the whistleblower, claiming that he or she is merely a troublemaker, a malcontent with an ax to grind or some sort of gripe with the organization. At the same time, the public and legal authorities may see the whistleblower as a hero who has taken great personal risk to reveal wrongdoing. Whistleblowers bring to light dangers and threats that would likely remain unknown and go unaddressed were it not for their willingness to speak out and risk their personal well-being.

As will become clear in this chapter, the debate about whistleblowing is in many respects a debate about what employees, and especially engineer-employees, owe their employers in the way of loyalty. Do engineers have an obligation of loyalty to their employers as the codes of ethics suggest? What can employers rightfully demand of their employees in the name of loyalty? Since whistleblowers often pay high personal prices for their actions, how far can or should the public expect engineers to go to protect safety, health, and welfare?

WHAT IS WHISTLEBLOWING?

The term *whistleblowing* originated in the practice of English policemen who blew a whistle when they observed a crime happening.[5] This alerted other police officers and the public to the fact that a crime was taking place. Although the parallel between the literal blowing of a whistle and the incidents referred to today as whistleblowing makes sense, formal definitions of whistleblowing are somewhat contentious. Consider the definition provided by Mike Martin and Roland Schinzinger: "Whistleblowing occurs when an employee or former employee conveys information about a significant moral problem to someone in a position to take action on the problem, and does so outside approved organizational channels (or against strong pressure)."[6]

This definition draws attention to four elements. The first is that acts of whistleblowing involve disclosure of something immoral. This may seem too obvious to mention, but it means that some public disclosures of information about organizations, even if unattractive to the organization, do not rise to the level of whistleblowing. For example, making information about a company's financial transactions or hiring practices public may be embarrassing to the company, but if they are not illegal or immoral, this is not whistleblowing. The information revealed by a whistleblower must involve wrongdoing—that is, a violation of law or well-established moral norm.

A second element of the definition is that whistleblowers are insiders. They are employees or former employees of the organizations that they accuse of wrongdoing. This element makes whistleblowers different from others who report organizational wrongdoing but have no formal employment relationship with the organization; for example, journalists are not whistleblowers when they write stories that expose organizational wrongdoing. The inclusion of former employees in the definition might be surprising, but former employees are considered to have obligations to their former employers, so they are considered insiders.

The third element is that whistleblowing involves going outside approved workplace channels. The idea here is that an employee has not blown the whistle if she simply reports a problem or a suspicion of wrongdoing to a supervisor or company ombudsman or uses some other formal process to express concerns. Such actions are understood to be an ordinary part of the activities of the organization. By contrast, whistleblowers are those who circumvent ordinary channels; they report their concerns "outside approved organizational channels."

This element of Martin and Schinzinger's definition diverges from others that distinguish between *internal* and *external* whistleblowers. Using this distinction, those who use ordinary channels of authority or circumvent hierarchical chains of command but stay within the organization are still whistleblowers; they are referred to as *internal* whistleblowers. Those who go outside the organization to report wrongdoing are referred to as *external* whistleblowers. Of course, an internal whistleblower might, at a later date, become an external whistleblower.

One reason for drawing a distinction between internal and external whistleblowers is that going outside the organization seems a greater act of disloyalty than staying inside, even if one circumvents ordinary reporting lines within the organization. For example, an engineer who observes his supervisor taking a bribe and reports this to top management does not seem to be disloyal to his employer even though he may be putting himself at risk— suppose that top managers are part of the bribery scheme and want the engineer to keep quiet. Nevertheless, the engineer could not be considered disloyal to his employer because in conveying the information to top management, he gives the organization an opportunity to address the problem quietly and without anyone outside the organization knowing. Or, he may give the organization the opportunity to make it public, thus controlling the story. Hence, although some might say that those who report their concerns inside an organization are not real whistleblowers, the notion of internal whistleblowing has the virtue of acknowledging that such acts do involve personal risk even though the risk is not as great as going outside the organization.

Unfortunately, the internal-external distinction is not as sharp as we might hope when it comes to real cases. Although Walter Tamosaitis is widely considered a whistleblower, the internal-external distinction does not easily apply to his employment situation. Initially, Tamosaitis reported his concerns to Bechtel, a company that was the client of the company that paid his salary (URS). Bechtel rejected his concerns and fired him, but being fired from Bechtel did not mean losing his job with URS. Tamosaitis continued to work for URS. So, at that point, it was hard to say whether he was a whistleblower—even an internal whistleblower— since Bechtel was not his primary employer. When he reported his concerns to the Defense Nuclear Facilities Safety Board, he clearly became a whistleblower. However, the situation is complicated because at that time he was still employed by URS though he was blowing the whistle on Bechtel. Hence, Tamosaitis's behavior does not fit neatly into the internal-external distinction. The final element of Martin and Schinzinger's definition is that the disclosure of wrongdoing must be made to someone or some group that can address the problem. This requirement is also tricky in that it seems to suggest that merely making some sort of wrongdoing public is not enough to make one a whistleblower. To be a whistleblower one must convey information "to someone in a position to take action on the problem." Salvador Castro's case is interesting in this context. Although public accounts of the case are scarce, it appears that Castro went only to his supervisor with his concerns and did not circumvent ordinary channels of authority. His concerns became public only because he sued his former employer for unjustly firing him. As a result of the suit, the problem with the incubators became public and was addressed by appropriate authorities. So, it is not clear whether Salvador Castro would be considered a whistleblower on Martin and Schinzinger's definition since he didn't go directly and externally to those who could do something about the problem. Walter Tamosaitis, on the other hand, initially went to those inside the organization who could do something about it and ultimately went outside to Congress, which could certainly do something about it.

As the Castro case suggests, whistleblowers may not always set out to be whistleblowers; they may not intentionally target an authority or agency that can address the problem. Some just go to the media with the idea that bringing the wrongdoing to public attention will result in its being addressed. In this respect, it seems too restrictive to make it a requirement of the definition of whistleblowing that one must convey information to "someone in a position to take action on the problem." Rather, it seems enough to say that whistleblowers convey information to those who can take action on the problem or they make information public in the hope that appropriate authorities or public pressure will lead to the problem being addressed.

The point of this analysis of Martin and Schinzinger's definition of whistleblowing is to suggest that the concept of whistleblowing is not as well understood as one might expect. In real-world situations, definitions apply only roughly. Nevertheless, even a rough definition of whistleblowing, such as that provided by Martin and Schinzinger, is helpful insofar as it identifies elements to look for when thinking about whistleblowing and especially when considering whether particular acts are justified. What is the relationship between the whistleblower and the organization accused of wrongdoing? What is the nature of the wrongdoing that is being reported? Did the whistleblower tell appropriate authorities inside the organization before going outside? Did the whistleblower try to get the problem fixed before going public?

Most engineering professional codes of ethics specify that engineers should be loyal to their employers, and employers often accuse whistleblowers of being disloyal. Hence, to understand whistleblowing and when it is justified, we should more carefully consider the obligation of loyalty to employer.

LOYALTY TO EMPLOYER

Loyalty is often seen as a virtue, a character trait to be admired in those who exhibit it. Individuals who are loyal are praised and thought to be good because they stand by their friends, family, or country in hard times as well as good. For example, loyalty is seen as central to friendship relationships. Being a friend involves doing what is in the friend's interest even when doing so may be burdensome. If you shrug off a friendship when it gets to be too much trouble, then it seems you were never really friends.

Of course, loyalty is generally understood to have limits, and there is a great deal of uncertainty about what one is obligated to do in the name of loyalty and when the obligation should be overridden. For example, suppose your friend is accused of committing a heinous crime and you are asked to help in the criminal investigation. Does loyalty require that you refuse to cooperate? Should you refuse to cooperate even if you believe that your friend might be guilty? In such a situation loyalty may require certain kinds of behavior, for example, doing what you can to ensure that your friend is treated fairly, but not others, for example, lying to the police.

Because loyalty is seen as a virtue, it might be assumed that being loyal to one's employer is a good thing or, at least, not to be sluffed off lightly. Most employees are highly dependent on their employers for income, so they wouldn't want to be perceived as disloyal. However, the appearance of loyalty and loyalty are not the same thing, so the question whether loyalty is even appropriate in the employer-employee relationship is important.

Traditionally, the employer-employee relationship has been thought of loosely as contractual. Each party provides something to the other and receives something in return. The employee provides labor and expertise and receives compensation as well as opportunities to develop and acquire new knowledge and experience in return.

In addition to this informal contractual arrangement, engineers often sign a formal employment contract specifying what the engineer employee will be required to do and what he or she must refrain from doing, such as not revealing trade secrets or not working for companies in the same industry for six months after termination. Typically, employment contracts also include provisions about patent rights on the employee's inventions. Such contracts are legally binding unless they can be shown to be unreasonable.

Whatever the contractual arrangement, an enormous body of law, both state and federal statutes and case law, also governs aspects of the employer-employee relationship. There are laws protecting employees and job candidates from discrimination and from infringement on their legal rights. For example, employers can't require that employees do anything illegal, and they can't constrain activities outside of work such as engaging in political activity or voting, even when such activities are against the interests of the employer.

Despite the complex nature of the relationship, the dominant idea governing employer-employee relationships is the doctrine of employment at will. According to this doctrine, employers have the right to hire and fire at will. This doctrine recognizes that employers should be able to make hiring and firing decisions in the best interests of their companies. This doctrine is limited only when laws specify otherwise, as in the case of laws against discrimination. So, for example, even though employers have the right to hire and fire at will, they are not allowed to discriminate—that is, they can't hire or fire on the basis of race.

In the context of whistleblowing—because whistleblowing is often seen as an act of disloyalty—the big question is: What do employees owe their employers *in the name of loyalty*, that is, beyond contractual obligations? To be sure, employers want their employees to believe that they should be loyal and to be loyal, but it is quite another matter to show that employees owe their employers something more or different than what they owe according to the employment agreement.

Skeptics argue that the idea of employees owing loyalty to their employers is a myth perpetuated by employers because it serves their interests. Employers want employees to believe they have an obligation of loyalty because employees who feel a sense of loyalty are more likely to be motivated to perform well. Employers also argue that they need the loyalty of their employees because they must rely on them to serve their interests.

In recent years this skepticism has taken hold and the idea that employees owe anything by way of loyalty to their employers has been seriously challenged. For example, Juan Elegido writes: "It has become commonplace that the old implied employment contract under which employers offered employment for life in return for the employees' undivided attention and devotion is dead. . . . Supposedly, modern economic conditions put a premium on employer flexibility and employee mobility and have rendered that implied contract unviable."[7] Elegido claims that because employees can no longer expect lifelong employment, employers can no longer claim that employees owe them loyalty.

When loyalty is taken out of the picture, employer-employee relationships are merely instrumental relationships. Each party uses the other for its own ends; each pursues its own self-interest. There is mutual benefit in such relationships in the sense that each gets something it wants. In this way of thinking about employer-employee relationships, one might worry that they violate the categorical imperative (introduced in chapter 3). However, remember that the categorical imperative is not violated if individuals are not treated *merely* as means. If employers and employees are honest with one another about what they are giving and receiving and if both parties refrain from coercion and exploitation, then the categorical imperative is not violated.

Framing employer-employee relationships as instrumental relationships is in harmony with the doctrine of employment at will—that is, employers are entitled to hire and fire as they wish and without cause (as long as they do not violate any laws). It is also in harmony with the idea that employees who blow the whistle are not being disloyal since they have no obligation of loyalty. Of course, both parties are legally required to refrain from certain actions, for example, employers can't violate whistleblower protection laws and employees can't violate the terms of their employment contracts. Nevertheless, if employer-employee relationships are instrumental, loyalty seemingly has no place in the relationship.

There may, however, be exceptions to this general position: loyalty to an employer may make sense in some circumstances. In other words, even though loyalty may not be an intrinsic part of every employment relationship, it may be owed in some relationships. Imagine, for example, that a company hired an employee who was just released from prison. There is no legal requirement to hire ex-convicts, but this employer took a chance on an employee who would otherwise have a difficult time finding a job. Imagine further that when the company went into a financial crisis, the employer took a pay cut rather than fire the ex-convict and others. Even further, imagine that the employee had a serious illness for some time and the employer allowed the employee to take much more sick leave (with pay) than the employment contract called for.

This employer's treatment of this employee goes far beyond what is legally or even morally required. The employer is seemingly not acting as if the relationship were merely instrumental. An employee treated in this way might rightly feel a special loyalty to this employer. Therefore, when the well-treated employee is approached by a law enforcement agency and asked to spy on the employer—because the employer is suspected of a crime—the employee might rightfully refuse to serve as a spy. The case would be more difficult to decide were there a preponderance of evidence that the good employer was engaged in harmful and unlawful behavior, such as selling unsafe products, money laundering, or engaging in child pornography. However, the point is that an obligation of loyalty would have developed from the extra things the employer did and not merely from the fact of an employer-employee relationship.

In contrast to the good employer case, imagine the employer who is entirely unworthy of loyalty. We can even imagine an employer who fulfills all the informal and formal terms of the employment contract, adhering strictly to the letter of the contract. This employer might pay employees on time, give them the specified vacation and sick days, and promote employees in a timely fashion and as deserved. Yet this employer does nothing above and beyond what is required. In essence, the employer treats employees as if they have an instrumental relationship. The employer gives employees their due and expects them to give what is owed and nothing more. In such a case, an employee may not owe anything to the employer *in the name of loyalty*. This is not to say that an employee should act contrary to the interests of this employer, but only that employees would owe nothing above and beyond what is required in the formal and/or informal employment agreement.

In the first situation, an obligation of loyalty developed from a series of actions that were above and beyond the contractual employment relationship. In the second situation, the relationship remained purely formal and contractual and loyalty is inappropriate. Of course, as already suggested, even in the case where loyalty is owed, there are limits to what an employee owes. We can imagine situations in which despite extra good treatment by the employer—above and beyond what is required—an employee is justified in cooperating with legal authorities investigating the employer for illegal behavior. Indeed, it is not hard to imagine that an employee might be justified in blowing the whistle on an employer who had been extremely good to the employee because the employee finds that this wonderful employer is engaged in some terrible, illegal, and dangerous activity.

This takes us to the justification of whistleblowing, for it is precisely when an employee sees that his or her employer is engaged in some dangerous, illegal, or immoral activity that the employee must decide whether to remain silent, try internal channels, or go outside the organization to get the issue addressed. Loyalty may or may not come into play here, but such decisions are rarely easy.

JUSTIFYING ACTS OF WHISTLEBLOWING

As mentioned earlier, whistleblowing can be very costly to whistleblowers as well as to those who are accused. Moreover, acts of whistleblowing can harm others who are not directly involved in the wrongdoing. In the Snowden case, the release of classified documents was said to have been harmful to the security of the United States and to the individuals who had been working as US spies whose names were released in the documents that Snowden stole. So, whether acts of whistleblowing involve disloyalty or not, they need justification because they can cause harm.

Whistleblowing has received attention in many contexts, but one of the first cases in which engineers were the center of focus was one that raised the question not just whether engineers might be justified in blowing the whistle but whether they had an obligation to blow the whistle. This was the topic of a seminal article in the field of engineering ethics by Richard De George. The article, which appeared in 1981, is especially interesting because it recognizes the special expertise of engineers and how that expertise is deployed in complex bureaucratic organizations. This is the case of the Pinto, a subcompact car made by the Ford Motor Company and sold during 1971–80. Ford engineers reported a problem with the Pinto to their superiors, but after top management decided not to fix it, the engineers did nothing more. The problem concerned the placement of the gas tank: when a Pinto was hit from the rear by another car traveling at low speeds, the gas tank would explode, killing the driver and passengers. In the legal case that ensued from an accident that caused the death of three young women, the question arose whether the engineers who had reported their concerns to managers should have blown the whistle on Ford when the company took no action.

In the blame game that followed revelations of the problem, the engineers came under scrutiny because they had deferred to management instead of persisting in their attempts to get the problem fixed. In their defense, De George argued that the engineers had no obligation to blow the whistle, and in his analysis of the case, he developed an account of when whistleblowing is justified, that is, when it is morally permissible, obligatory, and not obligatory to blow the whistle.

According to De George, whistleblowing is morally permissible when:

1. The harm that will be done to the public is serious and considerable;
2. The whistleblower makes their concerns known to their superiors; and
3. The whistleblower exhausts the channels available within the corporation, including going to the board of directors.[8]

The need for the first condition has already been discussed. Because whistleblowing has the potential to harm those who are accused as well as the whistleblower and others, the harm to be avoided or addressed must be significant enough to counterbalance these potential harms.

The second and third conditions specify that whistleblowers must try internal channels to have their concerns addressed before they go public. This is one of the key and controversial conditions of De George's account. Whistleblowers are only justified in blowing the whistle, according to De George, after they have made their concerns known to their superiors and exhausted internal channels to get the problem addressed. In this respect, De George is focused only on external whistleblowers, and he claims, in effect, that external whistleblowers must be internal whistleblowers first.

One reason that might be given for requiring engineers to make their concerns known to their superiors before going public is the obligation of loyalty to employer. Although we challenged this idea earlier, we also suggested that in certain circumstances employees may have an obligation of loyalty. Moreover, there are other reasons to try internal channels first. One is that in many circumstances one only has suspicions of wrongdoing, and it is better to find out if those suspicions are accurate by talking with someone inside the organization. Also, although it doesn't follow directly from the categorical imperative, it does seem that respect for other persons (or entities) would entail giving them the benefit of the doubt and, hence, the opportunity to explain what appears to be wrongdoing and/or giving them the chance to fix a problem without being exposed to public scrutiny. So, there is something to be said for trying internal channels before going public with one's concerns about wrongdoing in the organization in which one works.

In any case, none of these are the reason De George gives for requiring that engineers try internal channels first. He argues that engineers have only a limited perspective on what is in the best interests of a company and that their job is to provide their expertise and then leave it to management to integrate engineering judgments with other factors that must be considered in corporate decisions. In other words, engineers have one role to play in the corporation, and managers and executives have a different role. Management's role is to receive input from engineers and to put that input together with a variety of other considerations. Although engineers are primarily concerned with safety, safety is not the only concern of a corporation. Managers and executives must bring together safety considerations with considerations of cost, deadlines, regulatory requirements, market competition, customer wishes, client needs, and more. Thus, engineers should not second-guess the overall good of the company.

In the Pinto case, the question was whether Ford Motor Company had put an unsafe vehicle on the road. During the trial, it became clear that safety is not always a cut and dried matter. Indeed, even though the Pinto had a serious problem that could have been fixed with the addition of a relatively inexpensive part, Ford was found not guilty on grounds that the Pinto was no less safe than other cars in the same size and price category.

De George's concern about engineers second-guessing management has to do with recognizing that decisions made by employers—even those involving safety—may involve considering much more than engineers take into account. He suggests that engineers do not and are not required to see the big picture and therefore that they may not fully understand the significance of the behavior they observe as wrongdoing or even illegality. This leads De George to recognize that engineers can make mistakes with their accusations. More important, he recognizes that there may be genuine disagreements among engineers and between engineers and managers on how safety issues should be handled. Engineers and management can disagree about how dangerous an organization's behavior is or whether the organization is complying with the law or whether the organization is engaged in wrongdoing.

This is why De George distinguishes circumstances in which it is permissible to blow the whistle from those in which it is obligatory. He is reluctant to make whistleblowing obligatory when conditions 1–3 are met. An engineer may blow the whistle when these conditions are met, but it is not obligatory because engineers should also recognize that management balances multiple factors along with safety. On the other hand, according to De George, if two further conditions are met, engineers are not just permitted to but have an obligation to blow the whistle. These two conditions are:

4. The whistleblower has documented evidence that would convince a reasonable, impartial observer that the whistleblower's view of the situation is correct and what the organization is doing is wrong; and
5. There is strong evidence that making the information public will in fact prevent the threatened serious harm.[9]

Inclusion of these two conditions is best explained by returning to De George's concern with genuine disagreements between individuals about the safety of something going on in the company.

He is certainly right to be worried that some individuals may jump too quickly to a conclusion about wrongdoing without knowing all the facts or relevant considerations. He is also trying to make the case for when engineers are obligated to blow the whistle, and this must involve a fairly high bar so that engineers are not continuously challenging their bosses and are not overly burdened by the obligation.

On De George's account, if an engineer is in a situation in which all five conditions are met, then the engineer has an obligation to blow the whistle and would be blameworthy if he or she did nothing. By contrast, if an engineer is in a situation in which only conditions 1–3 are met, although it would not be wrong for the engineer to blow the whistle, the engineer has no obligation to do so. That is, the engineer would not be blameworthy for refraining from external whistleblowing.

Clearly, De George sees whistleblowing as costly and therefore in need of justification; he also sees it as fraught with the potential to be misguided. Not surprisingly, then, he concludes with a discussion of changes that might be made in organizations, law, and professional societies so that engineers' concerns about safety can be addressed in ways that avoid the need to blow the whistle.

De George's analysis of whistleblowing brings to light many of the central issues a potential whistleblower might face: How significant is and how sure am I about the harm or wrongdoing? Do I have enough evidence? Can I get the problem fixed without going public? Am I sure that the problem cannot or will not be addressed internally? Of course, in real-world situations, these issues are complicated or aren't always clear. For example, the requirement that whistleblowers must exhaust internal channels before going public is problematic in situations in which trying internal channels will be dangerous for the potential whistleblower, may lead to a retaliation, or may result in crucial evidence being destroyed.

Some complexities of real-world whistleblowing were discussed in relation to Salvador Castro and Walter Tomasaitis. Others will become apparent as we discuss the case of Edward Snowden.

THE CONTROVERSIAL CASE OF EDWARD SNOWDEN

As explained earlier, in 2013 Edward Snowden gave classified documents to journalists showing that the NSA was engaged in unauthorized surveillance of American citizens. Among other things, the NSA was collecting phone records of citizens who had done nothing wrong. After releasing the documents, Snowden fled the United States, going initially to Hong Kong and finally to Russia. Were he to return to the United States, he would be immediately arrested for violating the Espionage Act. He has been charged specifically with unauthorized communication of national defense information and willful communication of classified communications intelligence information to an unauthorized person.

Snowden's behavior continues to be hotly debated. Discussion often focuses on whether he should be pardoned so that he can return to the United States without having to go to jail. For example, in January 2014, the *New York Times* opined: "Considering the enormous value of the information he has revealed, and the abuses he has exposed, Mr. Snowden deserves better than a life of permanent exile, fear and flight. He may have committed a crime to do so, but he has done his country a great service."[10]

On the other hand, in 2016, when the House of Representatives Permanent Select Committee on Intelligence issued its report, *Review of the Unauthorized Disclosures of Former National Security Agency Contractor Edward Snowden,* former chairman Devin Nunes said, "Edward Snowden is no hero—he's a traitor who willfully betrayed his colleagues and his country. He put our service members and the American people at risk after perceived slights by his superiors."[11]

Those who see Snowden as a hero tend to emphasize NSA's wrongdoing and the wrong that was done to American citizens. Those who see him as a traitor tend to emphasize the violation of law and harm that Snowden may have done by releasing classified information.

If we use De George's account to evaluate Snowden's behavior, the distinction between permissible and obligatory whistleblowing is a helpful starting place. Those who defend Snowden do not claim that he had an obligation to do what he did. Rather, they seem to suggest that even though he committed a crime, Snowden was in a situation in which he was justified in blowing the whistle. In this respect, the claim seems to be that his actions were morally permissible.

It is, nevertheless, worth noting that Snowden's circumstances seem to fit De George's fourth and fifth conditions. That is, based on what we now know, Snowden had evidence that would convince "a reasonable, impartial observer" that his view of the situation was correct (condition 4). We know this because the public accepted the evidence that Snowden provided of the NSA's wrongdoing; the agency's behavior exceeded its authority. Moreover, Snowden had good reason to believe that making the classified documents public would either stop the harm from continuing or, at a minimum, lead to a public discussion and put citizens on alert as to what the NSA was doing (condition 5). Again, we know that this condition was met because that is what happened—making the classified documents available led to public discussion and ended the NSA's activities. This means that if Snowden's circumstances also met conditions 1–3, he would have had an obligation to blow the whistle.

So, the key question is whether conditions 1–3 were met. There seems little question that condition 1 was met; the harm that he observed being done to the public was serious and considerable. What about conditions 2 and 3? From the accounts that are available, the question whether or to what extent Snowden made his concerns known to his superiors (condition 2) and whether he could have or should have exhausted all internal channels of authority (condition 3) are unclear and in dispute.

According to interviews with Snowden, he did try to get the matter addressed internally. For example, a 2013 *Washington Post* piece reports that Snowden "brought his misgivings to two superiors in the NSA's Technology Directorate and two more in the NSA Threat Operations Center's regional base in Hawaii." However, Snowden's claim that he brought his concerns to the attention of superiors is challenged by an NSA spokesperson who reported in 2014 that "after extensive investigation, including interviews with his former NSA supervisors and co-workers, we have not found any evidence to support Mr. Snowden's contention that he brought these matters to anyone's attention."[12]

In Snowden's defense, some argue that NSA's surveillance was at such high levels of authority that Snowden could not have succeeded in changing NSA practices—that is, even if he had convinced his superiors, as members of the intelligence community, they would never have challenged the NSA. His defenders argue further that if Snowden had tried to get the problem addressed, he would have been fired, which means that the problem would have gone unaddressed. In discussion of what he should have done, some say that he should have gone to members of Congress. But it is unclear how his efforts would have been received, and certainly if he had gone to Congress, the NSA would not have been happy with him and might have retaliated in any number of ways, including discrediting him and terminating his employment.

Snowden's situation points to a problem with condition 3. Requiring whistleblowers to exhaust internal channels before going public may sometimes counteract the point of whistleblowing and make it impossible for the public to find out about wrongdoing. In some situations, as in Snowden's, a whistleblower's attempts to use internal channels gives wrongdoers the opportunity to destroy evidence and discredit the potential whistleblower. For this reason it seems that in some cases whistleblowers may be justified in blowing the whistle even though they have not exhausted all internal channels.

To be sure, trying internal channels seems a good strategy for potential whistleblowers for several reasons. For one, potential whistleblowers often suspect wrongdoing but are uncertain and lack evidence. Going to someone inside the organization can clarify the situation. Potential whistleblowers might find that their suspicions were unfounded or that the problem is already being addressed. In this respect, trying internal channels can minimize harm. So, requiring that whistleblowers make their concerns known to their superiors (De George's condition 2) makes good utilitarian sense; it is better to address wrongdoing with as little harm as possible. On the other hand, requiring that whistleblowers exhaust channels within the corporation, including going to the board of directors (De George's condition 3), can be quite problematic in real-world situations.

Conditions 2 and 3 aside, the Snowden case points more broadly to the matter of whom a whistleblower should approach outside the organization and what a whistleblower should reveal to demonstrate that there is a problem. These details of a whistleblowing matter quite a lot. These two matters are somewhat intertwined in the Snowden case. On the first matter, one wonders whether Snowden was right to turn over the documents to journalists, and on the second matter, one wonders whether he did the right thing in turning over all the documents. Although we should always be careful about judging how someone behaves in challenging situations, one can't help but wonder whether Snowden could have been more selective in who he turned over the documents to and whether he could have turned over fewer documents. This is another way of asking whether Snowden could have established the case against the NSA while at the same time minimizing harm done to others. Of course, the number of documents (over 1.5 million) was such that being selective about the documents could have been practically impossible.

Although we will not pursue these matters further, the Snowden case illustrates that engineers may confront a daunting decision when they see wrongdoing in an organization in which they work. As with other ethical challenges, engineers' responses to such discoveries rarely call for a single or simple action. Generally, such situations require an iterative process that includes finding out more, identifying those who can be consulted, knowing what channels of communication are available, and then taking one step at a time depending on the response at each step. Each next step depends on what happened with the prior step.

CONCLUSION

Are whistleblowing engineers heroes or traitors? As explained at the onset, the question can't be answered in the general. Each case must be examined on its own, and it isn't always easy to determine. What is clear is that whistleblowers often do a great service to society and at great personal cost. The public needs whistleblowers; they are an important means by which we find out about wrongdoing that would otherwise be kept secret. However, as much as whistleblowers are needed, so are strategies to eliminate the need for whistleblowing. It is not good for employers or engineers when legitimate concerns go unheard and unaddressed and reach a point where an engineer feels the need to blow the whistle. Organizations have been developing strategies to do just this, for example, creating an ombudsman position to which employees can anonymously discuss their concerns and making anonymous hotlines available to employees so that they can report their concerns without getting into trouble or making trouble for the organization. Even more important is the strategy of cultivating organizational cultures in which engineers and other employees feel free to speak out and air their concerns without fear of retaliation. The importance of developing such cultures will be discussed further in chapter 6.

SUGGESTED FURTHER READING

Fidler, David P. *The Snowden Reader*. Bloomington: Indiana University Press, 2015.

Hoffman, W. Michael, and Mark S. Schwartz. "The Morality of Whistleblowing: A Commentary on Richard T. De George." *Journal of Business Ethics* 127, no. 4 (2015): 771–81.

Sakellariou, Nicholas, and Rania Milleron, eds. *Ethics, Politics, and Whistleblowing in Engineering*. Boca Raton, FL: CRC Press, 2018.

PART THREE

Engineers, Safety, and Social Responsibility

6 ARE ROTTEN APPLES OR ROTTEN BARRELS RESPONSIBLE FOR TECHNOLOGICAL MISHAPS?

W HEN bad things happen, people want to know why. We want to know what went wrong, who was at fault, and what will be done about it. We also want to know if there are lessons to be learned—lessons that can be used to prevent the same kind of thing from happening again. Our interest in understanding why bad things happen is especially compelling when it comes to technological disasters and incidents of large-scale wrongdoing.

Engineering ethics has been shaped by examination of cases in which the results were catastrophic in terms of loss of life, environmental degradation, or financial impact. These cases raise a host of ethical as well as technical questions. How could safety standards at a Union Carbide factory in Bhopal, India, come to be so lax that an explosion occurred, killing thousands of workers? Why did Ford Motor Company knowingly choose to do nothing about the dangerous placement of the gas tank on the Pinto? Why did NASA launch the *Challenger* with uncertainty about how the seals on the rocket boosters would behave in very cold temperatures? Answering such questions can provide important lessons for engineers and the organizations in which they work, lessons that have everything to do with safety.

In this chapter, we examine an age-old question: Is it rotten apples or rotten barrels that lead to bad things happening? In other words, is it the bad behavior of individuals or pernicious environments that lead to technological accidents and wrongdoing? As with other debates taken up in this book, the answer need not be one or the other. The question is posed as an either/or only so that different aspects of technological mishaps can be examined. As we will see, rotten apples and rotten barrels tend to go together—that is, technological accidents tend to occur when individuals (especially those at the top of an organization) make poor or even unethical decisions and some of their bad decisions involve creating environments that, in turn, make it more likely for individuals to behave badly. We will see this played out in several of the cases discussed below.

Traditionally individuals and their decisions have been the focus of attention in identifying the human causes of untoward events. Only recently has the importance of organizational environments come into focus. This new focus was especially notable for engineering in the aftermath of the *Challenger* disaster. On January 28, 1986, the NASA space shuttle *Challenger* broke apart seventy-three seconds into its flight, killing all seven crew members. Although there is a much longer story about what happened, the physical cause of the explosion was the failure of the O-ring seals used in the joints of one of the solid rocket boosters. The O-rings failed to hold due to the unusually cold weather at the time of the launch. The decision to launch was made despite concerns expressed by engineers over how the seals might function in cold temperatures.

Some years later Diane Vaughan published an analysis of the incident suggesting that it wasn't just poor decision-making that led to the failed launch but a culture at NASA in which certain risks had been normalized.[1] Vaughan referred to what had happened as the "normalization of deviance." She succinctly explained how this worked in an interview for ConsultingNewsLine in 2008:

> Initially, it appeared to be a case of individuals—NASA managers—under competitive pressure who violated rules about launching the shuttle in order to meet the launch schedule. It was the violation of the rules in pursuit of organization goals that made it seem like mi[s]conduct to me. But after getting deeper into the data, it turned

out the managers had not violated rules at all, but had actually conformed to all NASA requirements. After analysis I realized that people conformed to "other rules" than the regular procedures. They were conforming to the agency's need to meet schedules, engineering rules about how to make decisions about risk. These rules were about what were ac[c]eptable risks for the technologies of space flight. We discovered that they could set-up the rules that conformed to the basic engineering principles that allowed them to ac[c]ept more and more risk. So they established a social normalization of the deviance, meaning once they accepted the first technical anomaly, they continued to accept more and more with each launch. It was not deviant to them. In their view, they were conforming to engineering and organizational principles.[2]

In short, Vaughan's analysis suggested that an aspect of NASA's culture blinded the engineers to the significance of the data they were seeing. Individual engineers weren't necessarily rotten apples, they were in a barrel that shaped what they thought and did. To be sure, we can raise serious questions about why individuals succumb to the pressures of bad organizational cultures, and we can search for ways to train individuals to resist the pressures of bad organizational culture, but we also should acknowledge that organizational culture can have a powerful influence on individual behavior.

RESPONSIBILITY, CAUSALITY, BLAME, AND ACCOUNTABILITY

Before we explore the bad apples–bad barrels debate, it will be helpful to consider how attributions of causality and responsibility are made in complex technological disasters. This is not a simple matter, and it is important to keep the complexities in mind.

The Fukushima Daiichi Nuclear Power Plant Disaster

On March 11, 2011, a magnitude nine earthquake erupted off the coast of Japan and caused a forty-six-foot tsunami to hit the coast near the Fukushima Daiichi Nuclear Power Plant. The wave surpassed and destroyed the protective wall that had been built in anticipation of tsunamis. An estimated twenty thousand people were swept out to sea and hundreds of thousands of people were evacuated. Although the earthquake triggered a shutdown of the reactors at the plant, when the cascades of water reached the plant, the generators used to cool down the reactors were flooded. As a result, there were hydrogen explosions and fuel core melt-downs, resulting in the complete destruction of the plant and the release of nuclear radiation that immeasurably threatened human health. The site continues to be dangerously contaminated.[3]

After the disaster, fingers pointed in many directions. The extraordinary size of the tsunami was seen by some as the cause of the disaster; that is, the incident could be seen simply as a *natural* disaster. In that case, no one would be responsible. At the same time, questions arose as to whether the engineers who designed the wall had been at fault for not designing a higher wall. On that view, the short-sightedness of the engineers was seen to be the cause of the disaster. Contradicting this view, others argued that the magnitude of the tsunami was such an unlikely event that the engineers should not be faulted for not anticipating it. If one believes that the engineers made a reasonable decision about how high to build the wall, then the disaster looks more like a natural disaster. The difference in positions here has to do with a norm for building protective walls. In many cases there are regulations requiring engineers to build to a certain standard. When they adhere to the standard, they are not considered blameworthy, and when they fail to build to the standard, they are considered neg-ligent. However, even when there are standards and engineers adhere to the standards, those who set the standards may be blamed for setting the standards too low. Judgments as to what the standard should be or what level of probability engineers should consider in their designs are not simple or straightforward matters. Building for low-probability events is significantly more expensive than building exclusively for high-probability events.

The point is that identifying the cause of a technological mishap and who is responsible necessarily involves normative judgments as well as technical accounts. We need to know what happened physically as well as what norms of behavior are relevant to the human actors involved. We need to know whether or to what extent human actors failed to adhere to prevailing norms.

The role of norms is also evident in other aspects of the Fukushima disaster. The severity of the disaster could be blamed on the engineers who designed the plant. As mentioned, water reached the generators used to cool down the reactors, and this happened because the pumps used to keep water from the generators were flooded. Had the pumps functioned, the generators would have worked and the reactors could have been cooled down, and, arguably, the scale of the disaster would have been contained (though not avoided). Unfortunately, the design engineers had not anticipated that floodwaters would ever reach the plant, so they placed the pumps at a low elevation. Had the pumps been higher, floodwaters would not have reached them.

Whether we consider the engineers who designed the plant responsible for the failure of the pumps depends on whether we believe that the engineers should have anticipated floods reaching the plant. This is a normative judgment not unlike the judgment about the height of the protective wall. Both judgments have to do with the probability of something happening.

Other analyses of the Fukushima disaster focus on decisions made inside the plant as the crisis unfolded. Apparently, the engineers who responded inside the plant did not follow appropriate procedures. Again, if they had adhered to the norms for handling situations of the kind they confronted, they would not have been held responsible. Instead, however, their failure to do what was expected contributed to the harm done, and they were seen as a factor in causing the damage.

The Fukushima case shows that identifying the cause of a disaster is a complex, normative process and not just a matter of tracing the physical sequence of events. What is considered the cause and who is responsible depends on standards that apply to machines and devices as well as norms for how people, especially engineers, should behave. To be sure, norms and standards vary with the technology: the standards for building nuclear power plants are different from those for space shuttles, automobiles, and computers. The point is, nevertheless, that norms come into play in the evaluation of all technological disasters and all attributions of responsibility.

Sorting Out Responsibility

The physical and normative analyses identifying the cause or causes of an incident are used to attribute responsibility to individuals and organizations. In sorting out accidents, certain terms are often used loosely and interchangeably: *responsibility, accountability, blameworthy, liable. Responsibility* is the broadest term, and when we ask who or what is responsible for an event, we can mean any number of things. Sometimes we are asking what or who is the cause of the event; other times we want to know who is accountable—that is, who we should go to for an explanation of what happened and why; other times we want to know who is blameworthy; and yet other times we want to know who is the appropriate person or entity to compensate those who were harmed. Although there are cases in which these different matters all point to the same person or organizational unit, it is important to keep in mind that accountability, blameworthiness, and liability are somewhat different notions and can apply to different people in the same incident.

Typically, the accountable person or entity is the head of the organization in which the accident or wrongdoing occurred. However, this person or organization may be quite distant from the activities leading up to the event. For example, in the *Challenger* disaster, President Ronald Reagan was ultimately accountable. NASA is an independent agency of the executive branch of the federal government, so the public expected the president to answer for what had happened. Acknowledging his accountability, President Reagan appointed a commission to report on what happened soon after the accident.[4] However, no one explicitly blames President Reagan or holds him personally liable for the accident. There were critics who suggested that NASA was under greater pressure to launch because President Reagan had raised public expectations about the launch and a failure to launch might have been viewed as a bad mark for his presidency. Nevertheless, President Reagan was considered accountable but not blameworthy or liable for the actual accident.

Of course, when a complex technological accident occurs, more than one person is usually accountable. Even though President Reagan was ultimately accountable for the *Challenger* disaster, the head of NASA was also accountable, and so were many other decision makers at various places in the NASA organization.

Liability is different from *accountability* and *blameworthiness* in that it refers to the obligation to compensate for the harm that was done. An individual or organization may be liable to pay compensation or to pay a fine, and sometimes we say that individuals are liable to punishment. In many technological mishaps the liable party will be the organization in which the mishap occurred, but generally in such cases, individuals who are at the head of or in leadership roles in organizations are not held personally liable. Of course, if they did something illegal, they will be criminally liable, but that is another matter.

The search for who is to blame is the most complicated because deciding who is blameworthy is intertwined with causality and the norms for the situation. We saw this in discussing the Fukushima Daiichi case, where the engineers who designed the protective wall and those who designed the plant were blamed for having not anticipated a tsunami of the size that occurred. In the cases discussed below, the question of who or what is to blame takes us directly to the rotten apples–rotten barrel debate. Each case is extremely complex, and we will only highlight certain features—features that drew public attention in the search for an explanation of what happened and why. Those who wish to pursue the details of the cases further will find a list of key reports at the end of the chapter.

BAD APPLES, BAD BARREL IN RECENT CASES

We begin with a case that points to individual engineers as the rotten apples. In the Kansas City Hyatt Regency Hotel collapse, two structural engineers were found to be negligent in their overseeing of a huge construction project that collapsed, killing more than a hundred people and injuring many more. The second case introduces the idea that an organizational environment can be a major contributor to bad things happening. The *Columbia* space shuttle accident harks back to the *Challenger* disaster and the importance of creating a safety culture. The General Motors ignition switch case takes us further into organizational culture and how it might lead engineers and an entire company to fail to identify and respond to a serious safety issue. Many people lost their lives or were harmed before GM recognized and did something about a faulty ignition switch in several GM models. The Volkswagen emission fraud case is an example of outright, intentional wrongdoing; it couldn't be called an accident. It was a disaster for the company, its customers, and the environment and could be seen as a case of both rotten apples and a rotten barrel.

The Kansas City Hyatt Regency Hotel Collapse: Negligence of Engineers

Although it took some time, investigations after the Kansas City Hyatt Regency Hotel collapse indicated that the engineers who oversaw the project were to blame. Hence, it seems a clear case of rotten apples. Of course, we have to be careful in using the term *rotten apples* because no one claims that the engineers set out to produce a collapse. Saying that it was a case of rotten apples simply means that it was the *behavior* of particular individuals that led to (caused) the accident.

On July 17, 1981, between fifteen hundred and two thousand people were dancing in the atrium of the Kansas City Hyatt Regency Hotel. The dance was a weekly event sponsored by the hotel. As people danced, two of the suspended walkways inside the atrium collapsed. One hundred and fourteen people were killed and almost two hundred more were injured. The National Bureau of Standards report described the incident as "the most devastating structural collapse ever to take place in the United States" in terms of loss of life and injuries.[5]

Sarah Pfatteicher's explanation of what happened succinctly describes the physical causal sequence of events:

> Investigators . . . traced the walkway collapse to poor connections between the walkways and the hanger rods from which they were suspended. The design of the skywalks had called for both the second- and fourth-floor walks to hang from a single set of rods suspended from the lobby ceiling and running through both walkways. But the skywalks were not built according to the original plan. Instead, the fourth-floor walk hung from rods connected to the ceiling, and the second-floor walk hung from an additional set of rods connected to the fourth-floor walk. The change effectively doubled the load on the fourth-floor walkway connection. Disaster was almost assured.[6]

Although this aspect of the collapse was clear, it left the question of how such a mistake could be made. Who was at fault? As the search for answers proceeded, fingers were pointed in many directions: Was it the conceptual design of the building? The construction company? The steel rod fabricators? Although some details are still somewhat unclear, the structural engineer and his assistant were found to be blameworthy for not understanding or catching the significance of the change from the original design. A grand jury charged with investigating whether any illegal actions had led to the collapse found "no evidence of wrong-doing on the part of the design professionals," but in a subsequent state of Missouri administrative hearing, the engineer of record (EOR) and his assistant were found guilty of gross negligence, misconduct, and unprofessional conduct.[7]

Accounts vary somewhat as to how the change in design was made. However, the EOR was accountable for the design and was expected to sign off on any changes made to the design during construction. Apparently, the assistant to the EOR reviewed the change in the rod arrangement and had the EOR sign off on the change, but it is unclear how carefully the EOR examined the change before signing off. Although it could be argued that the decision about how to suspend the walkways fell between the cracks, according to Pfatteicher on six separate occasions the assistant to the EOR had been asked about the implications of the design change, and each time he assured the inquirer that the change would not compromise the safety of the walkways.[8] In the Missouri administrative hearing, the judge found the assistant to the EOR guilty of "gross negligence for failing to monitor the manufacture of the box beam connectors that supported the walkways," and the EOR guilty of negligence for not adequately supervising the assistant.[9]

So, this case clearly points to the negligent behavior of individuals as the primary cause of the accident. Although organizational environment was not a theme in analyses of this incident, some have pointed the finger at an approach to construction that was fashionable at the time. Gregory Luth refers to this as the fast-track method of delivery and explains that many projects delivered by this method in which construction precedes design were "plagued by a lack of time and quality control."[10] This hints at the idea that the accident may not just have been the result of individual behavior; an organizational factor may have played a role in what happened.

The *Columbia* Space Shuttle Explosion: Organizational Culture

On January 16, 2003, the space shuttle *Columbia* lifted off for a sixteen-day science mission featuring microgravity experiments. As the spacecraft reentered the Earth's atmosphere on February 1, 2003, it exploded. All seven crew members were killed. NASA created the Columbia Accident Investigation Board (CAIB) to investigate the causes of the accident and on August 26, 2003, the CAIB issued its report. As with the report on the *Challenger* disaster, the new report raised serious issues about the organizational culture at NASA.

The physical cause of the explosion was relatively quickly identified as a piece of foam insulation that had broken off from the external fuel tank during launch, striking the left wing of the orbiter. (The orbiter is the airplane-shaped part of the shuttle that launches together with two solid rocket boosters and an external fuel tank; the orbiter houses the crew.) When the foam piece struck the orbiter's wing, it created a hole. The hole didn't affect the mission until reentry, when it allowed superheated air to enter the wing. This caused the wing to fail structurally, and this, in turn, created aerodynamic forces that led to the disintegration of the orbiter.

Harking back to our earlier discussion of the complexities of identifying *the* cause of an accident, the CAIB did not leave its analysis at the physical causes. The report is unambiguous in concluding that the accident was the result of technical *and* organizational failures. In fact, the CAIB gives a good deal of attention to the organizational causes and includes organizational changes in its lists of recommendations.[11]

In the context of our rotten apples–rotten barrels debate, the report does far less finger-pointing at individuals and puts more emphasis on the organizational environment. When the report was published, the media also seemed to pay more attention to the report's findings regarding organizational culture. This was likely because it appeared that NASA had not heeded the lessons of the *Challenger* disaster seventeen years earlier. As one *Washington Post* journalist explained:

> In 248 unsparing pages, the board not only identified the physical cause of the disaster but also the deep-seated managerial and cultural problems that, panel members found, were as much to blame as the chunk of foam that hit the shuttle's left wing 81.9 seconds after launch. These systemic problems are far more disturbing and will be far harder to fix than the insulating foam. Saddest of all is the board's conclusion that NASA failed to learn "the bitter lessons" of the 1986 *Challenger* explosion and fix the "broken safety culture" identified after *Challenger*."[12]

The report points to many dimensions of organizational culture, some having to do with the special risks associated with space travel and others more generalizable to all contexts in which large complex systems put human beings at risk.

The GM Ignition Switch Case: Failure to Act

In the General Motors ignition switch case, an organizational culture that seemed to work against safety was also targeted in the investigation of what went wrong. In this case, GM's board of directors ordered the investigation and appointed Anton

Valukas, a former US attorney for the Northern District of Illinois, to head up the investigation and report to the board. The report, often referred to as the Valukas Report, provides the following sketch of what happened:

> In the fall of 2002, . . . a GM engineer chose to use an ignition switch in certain cars that was so far below GM's own specifications that it failed to keep the car powered on in circumstances that drivers could encounter, resulting in moving stalls on the highway as well as loss of power on rough terrain a driver might confront moments before a crash. Problems with the switch's ability to keep the car powered on were known within GM's engineering ranks at the earliest stages of its production, although the circumstances in which the problems would occur were perceived to be rare. From the switch's inception to approximately 2006, various engineering groups and committees considered ways to resolve the problem. However, those individuals tasked with fixing the problem—sophisticated engineers with responsibility to provide consumers with safe and reliable automobiles—did not understand one of the most fundamental consequences of the switch failing and the car stalling: the airbags would not deploy. The failure of the switch meant that drivers were without airbag protection at the time they needed it most. This failure, combined with others documented below, led to devastating consequences: GM has identified at least 54 frontal-impact crashes, involving the deaths of more than a dozen individuals, in which the airbags did not deploy as a possible result of the faulty ignition switch.[13]

Ultimately, GM recalled 2.6 million cars and has paid more than two billion dollars to settle claims.[14]

GM initially put forward an account of what had happened that minimized the problem and suggested that few individuals had been involved. As more of the details were revealed, the narrative changed, revealing a long history of awareness of a problem with the ignition switch and inaction. Ultimately the inaction was connected to a dysfunctional corporate culture. Typically the CEO of a corporation is held accountable for accidents of this kind, and the CEO of GM at the time that the problem came to public attention had only recently been appointed. Mary Barra had been employed at GM when the company knew about the problem, but her role had been distant from the issue. So when it became public, she was not defensive; she acknowledged the company's accountability and openly discussed the problem and how it had been dealt with. She also pledged to change things.

Where organizational culture is suspected of being a problem, one of the first things examined is how the organization prioritizes cost and safety. The suspicion is that the organization may put cost ahead of safety. According to the Valukas Report, before the ignition switch problem, GM had a culture of prioritizing safety. However, in recent years safety had come into more heightened tension with cost because of massive financial troubles that GM had encountered. Staff had been severely cut, and cost became a dominant theme in decision making. The Valukas Report explains the situation this way:

> We have uncovered no evidence that any employee made an explicit trade-off between safety and cost. . . . That noted, we cannot conclude that the atmosphere of cost-cutting had no impact on the failure of GM to resolve these issues earlier. . . . GM was under tremendous cost pressure, and it was imposing tremendous pressure on its suppliers to cut costs. Engineers did not believe that they had extra funds to spend on product improvements. Staff was cut dramatically. It is not feasible for three to do a job as effectively as eight.[15]

Nevertheless, financial conditions were not the only problem. GM culture is singled out as especially problematic. The Valukas Report identifies four elements in GM culture that were relevant to the company's failure to act on the ignition switch. First, there was an atmosphere in which employees were reluctant to raise issues and concerns. This was exacerbated by GM's monetary problems because the financial problems made employees disinclined to put forward problems that would likely call for costly solutions. Second, employees described a culture in which something they called "the GM nod" was commonplace. This was the phenomenon in which everyone at a meeting would nod in agreement to do something, and yet no one had any intention of following through and no one would be explicitly tasked to do anything. The nod was "an empty gesture." As the Valukas Report explains: "We repeatedly heard from witnesses that they flagged the issue, proposed a solution, and the solution died in a committee or with some other ad hoc group exploring the issue. But determining the identity of any actual decision-maker was impenetrable. No single person owned any decision."[16]

Third, the environment did not encourage knowledge sharing. This lack of knowledge sharing contributed to a situation in which the engineers working on the ignition switch problem did not know that the car was designed so that the airbags would not deploy when the ignition switch was off. Ignorance of this connection is understood to be a major factor in explaining GM's inaction. Not knowing that airbags were turned off when the ignition went off, GM engineers had classified the ignition switch problem as one of "customer convenience" and had not given it the priority it deserved. The Valukas Report suggests that in an environment that encouraged knowledge sharing, the engineers more likely would have learned about the connection.

Fourth, and ironically, the culture at GM included a notion that no action should be taken on a technical matter until one figured out the root cause of the problem. Engineers continued to be baffled by the problem with the ignition switch, and because of this, they thought it would be wrong to act until they figured it out. This, coupled with their belief that the problem was merely a matter of customer convenience, meant that they did nothing.

So, the Valukas Report comes down strongly on the rotten barrel side of the apple–barrel debate. However, it is important to note that it does not let individual engineers off the hook. In fact, the report criticizes the engineers for their lack of diligence in addressing the ignition switch problem: "Engineers investigating the non-deployments and attempting to understand their cause were neither diligent nor incisive. The investigators failed to search for or obtain critical documents within GM's own files, or publicly available documents that helped non-GM personnel make the connection between the switch and airbag non-deployment."[17]

Furthermore, investigative attention focused on one specific GM engineer, the one who had approved a change in the ignition switch but did not change the part number of the switch. He failed to document the change, told no one about it, and later claimed not to remember making the change.[18] This was a violation of GM policy, and it meant that later on the engineers investigating the problem could not figure out why some cars behaved differently than others even though they seemingly had the same ignition switch.

The GM ignition switch case provides a vivid picture of engineers getting caught up in organizational cultures that are not conducive to safety and problem solving. The case illustrates the importance of creating environments in which engineers can address issues of safety effectively, but at the same time it shows that the behavior of individual engineers can also play a role in bad things happening. Safety is a matter of both good apples and good barrels.

The Volkswagen Case: Intentional Wrongdoing

We turn now to a very different sort of case, one in which the untoward event could not be called an accident. It is a case involving outright intentional wrongdoing in which corporate culture seemed to play an important role. In 2015, the Environmental Protection Agency (EPA) discovered that diesel engine Volkswagens sold in the United States contained a defeat device.[19] The device consisted of software that detected when a car was in testing mode, and when this happened, the performance of the car changed so that emissions were reduced and the car passed the test. Once a car was put back into normal, nontesting mode, it spewed out nitrogen oxides at forty times the EPA permissible levels. On September 18, 2015, the EPA issued to Volkswagen a notice of violation of the Clean Air Act. The cars in violation included Volkswagen, Audi, and Porsche models, all manufactured by Volkswagen AG (VW), a multinational company headquartered in Germany.

The impact of the discovery was catastrophic for Volkswagen. Although the EPA initially discovered the fraud, other countries soon found that the defeat device had been installed on vehicles sold in their countries. The fraud had taken place over a ten-year period with a half million vehicles at issue in the United States and eleven million worldwide.[20] VW faced criminal charges, fines, and a variety of lawsuits from agencies in different countries as well as directly from customers. Some of the legal proceedings are still in progress, but as of September 2017 one source calculated that the incident cost VW thirty billion dollars.[21]

The deception had an ironic element that compounded the outrage of those who had bought the vehicles. The cars had been marketed by VW as "eco-friendly," and at least one model had won a Green Car of the Year award.[22] Many customers had felt satisfaction in believing their cars were good for the environment; they were devastated to learn that their cars were spewing out pollutants that were not only bad for the environment but dangerous to human health.[23]

Because a defeat device was used to fool EPA, there was no question as to intent. It had to have been intentionally designed and installed. The device stood as evidence of the wrongdoing. There was no question as to VW's accountability. VW was responsible for explaining what happened, why, and how, and it was liable to pay fines and compensation. The big question was who exactly was at fault and why they had done what they did. In this case, no major report has been issued. Exactly who did what and why has been slowly coming to light as legal proceedings have taken place.

Immediately following the revelation of the fraud, company spokespersons insisted that neither executives at the top nor the company's board knew about the fraud. The suggestion seemed to be that they were not blameworthy. However, soon thereafter the CEO of the company, Martin Winterkorn, resigned. Winterkorn claimed that he was unaware of the defeat device. His resignation suggested, then, that he acknowledged being accountable even though he did not see himself as blameworthy. Denial of the fraud aside, in May 2018 the United States filed criminal charges against Winterkorn, accusing him of conspiring to cover up the fraud.[24]

Although there is no question about the company's accountability and liability, the search for who is to blame has focused both on individuals and on corporate culture. Early on fingers were pointed at the engineers, for engineers had to have designed and implemented the defeat device. Many people were astonished at the engineers' behavior. As one business ethics professor wrote: "It's shocking that the software engineers of Volkswagen over-looked and neglected their fiduciary responsibility as professionals. Professionals who have a semi-regulatory responsibility within the organization to ensure safety, in this case environmental safety, even when this is less efficient or economical."[25]

The first VW employee to be sent to prison for the emission fraud was an engineer. James Liang may not have been the mastermind of the fraud, but he was a key player in developing the software device. He plead guilty and in August 2017 was sentenced to forty months in prison.[26]

Nevertheless, although individuals were implicated in the case, media attention has focused on the environment at VW and officials in the company who created a culture in which the fraud could occur.[27] The culture inside VW was described as "confident, cut-throat and insular."[28] According to reports, enormous pressure was put on employees to perform (or be fired), and they were discouraged from reporting or openly discussing problems. For example, Leah McGrath Goodman writes that "people familiar with the company tell *Newsweek* that the unique corporate culture of Volkswagen . . . led to an environment in which employees live and work under a highly centralized hierarchy that expects them to perform, no matter what the demands."[29] In another description J. S. Nelson reports that "shareholder advocates and former employees of VW criticize the construction of a culture inside the company that 'discouraged open discussion of problems, creating a climate in which people may have been fearful of speaking up.' . . . Even VW's internal investigation into the fraud has been hampered 'by an ingrained fear' in the company's culture 'of delivering bad news to superiors.'" Nelson notes that Germany's *Der Spiegel* magazine has famously described VW's culture as "North Korea without labor camps."[30]

Importantly, in the media discussion, even when corporate culture is seen as problematic, blame for the culture is directed at individuals. The discussion of VW culture has pointed directly at top executives for creating the environment. In particular, Ferdinand Piëch and Martin Winterkorn are picked out as culprits because of their management styles. Jack Ewing and Graham Bowley report critics as saying that "after 20 years under Mr. Piëch and Mr. Winterkorn, Volkswagen had become a place where subordinates were fearful of contradicting their superiors and were afraid to admit failure."[31] And Geoffrey Smith and Roger Parloff report that "VW is driven by a ruthless, overweening culture. Under Ferdinand Piëch and his successors, the company was run like an empire, with overwhelming control vested in a few hands, marked by a high-octane mix of ambition and arrogance—and micromanagement—all set against a volatile backdrop of epic family power plays, liaisons, and blood feuds."[32]

So, even though the VW emission fraud case is about intentional wrongdoing, which is radically different from the GM ignition switch case, it also points to an intertwining of rotten apples and rotten barrels. In particular, the VW case suggests that bad apples create bad barrels, which in turn leads to more (subordinate) bad apples. The picture suggested by media reports is that the engineers were given the technical challenge of developing a diesel engine car that would meet EPA standards, and when they couldn't do this, they came up with the idea of the defeat device rather than tell higher-ups that they could not do what they had been asked to do.

If we ask what lessons about ethics can be learned from the case, the clearest lesson—perhaps too obvious to be worthy of mention—is: Don't break the law. This is a lesson for corporations and for engineers working in corporations. However, there is also another lesson suggested by the bad apples–bad barrel issue, perhaps less obvious, though far from complicated. This is a lesson for leaders and others who create and shape organizational culture. The lesson is that creating environments in which employees are comfortable bringing up problems and having them discussed openly can be critically important to success. When problems are brought to light, they can be addressed; when problems are not brought to the fore, discussed, and addressed, bad things can happen.

CONCLUSION

So, are rotten apples or rotten barrels responsible for techno-logical mishaps? As explained at the outset, the issue need not be taken as an either/or matter. Individual behavior and organiza-tional culture can both play a role in technological accidents and wrongdoing. In particular cases, one may play a larger role than in others. However, more often than not, rotten apples and rotten barrels seem to go together. In fact, they seem to be so intertwined that it is difficult to disentangle them. Organizational leaders have a powerful influence in shaping organizational culture, and that culture, in turn, influences how individuals behave. This means that figuring out who or what is responsible for technological mishaps is multidimensional.

The flip side of this is that ethical engineering requires both good apples and good barrels. Individuals and organizational cultures that do not give priority to safety and adherence to law can have devastating effects. This takes us back to chapter 1, where the concern was whether education could make individual engi-neers ethical. In this chapter, we see that ethical engineering is not just a matter of ethical engineers; it is also a matter of orga-nizational cultures that encourage and enable ethical behavior. If we want engineers and engineering to be ethical, chapter 1 sug-gests that we focus on teaching and socializing individual engi-neers (into good apples), and this chapter makes clear that we must also focus on creating good organizations (good barrels)—that is, organizations that promote and support ethical conduct.

SUGGESTED FURTHER READING

Challenger Space Shuttle Accident

US Presidential Commission on the Space Shuttle Challenger Accident. *Report to the President: Actions to Implement the Recommendations of the Presidential Commission on the Space Shuttle Challenger Accident*. National Aeronautics and Space Administration, July 14, 1986. https://history.nasa.gov/rog-ersrep/genindex.htm.

Columbia Space Shuttle Explosion

Report of the Columbia Accident Investigation Board. *Space Shuttle Columbia and Her Crew.* National Aeronautics and Space Administration, August 26, 2003. https://www.nasa.gov /columbia/home/CAIB_Vol1.html.

Smith, Marcia S. *NASA's Space Shuttle Columbia: Synopsis of the Report of the Columbia Accident Investigation Board.* CRS Report for Congress, September 2, 2003.

Fukushima Daiichi Nuclear Power Plant Failure

Aoki, M., and G. Rothwell. "A Comparative Institutional Analysis of the Fukushima Nuclear Disaster: Lessons and Policy Implications." *Energy Policy* 53 (2013): 240–47.

World Nuclear Association. "Fukushima Accident." Updated October 2018. http://www.world-nuclear.org/information-library /safety-and-security/safety-of-plants/fukushima-accident.aspx.

GM Ignition Switch Case

Valukas, Anton R. *Report to Board of Directors of General Motors Company Regarding Ignition Switch Recalls.* Jenner & Block, Detroit, May 29, 2014.

Kansas City Hyatt Regency Hotel Collapse

Luth, Gregory P. "Chronology and context of the Hyatt Regency collapse." *Journal of Performance of Constructed Facilities* 14, no. 2 (2000): 51–61.

Volkswagen Emission Fraud

Hotten, Russell. "Volkswagen: The Scandal Explained." *BBC News,* December 10, 2015. https://www.bbc.com/news/business -34324772.

Topham, Gwyn, et al. "The Volkswagen Emissions Scandal Explained." *Guardian,* September 23, 2015.

7   WILL AUTONOMOUS CARS EVER BE SAFE ENOUGH?

IN the last chapter, we took a backward-looking perspective in examining technological accidents that occurred in the recent past and asking who or what was responsible in each case. In this chapter, we take a forward-looking perspective, asking what can and should be done now to avoid accidents in the future. We focus on an emerging technology, autonomous cars. In looking backward in the last chapter, we noted the importance of existing standards and norms and their role in identifying who or what was responsible when accidents and wrongdoing occurred. We didn't worry about where the norms came from or how they were established.

In this chapter we turn our attention to norms and standards for the safety of autonomous cars. Cars with low levels of autonomy are now on the road, and enormous resources are being put into the development of fully autonomous cars. However, safety standards are not yet established, and there hasn't been a broad public discussion of what level of safety should be required. Moreover, it is not clear how standards will be established. In order to explore what level of safety can be or should be achieved in autonomous cars, this chapter considers the pros and cons of autonomous cars and asks how the safety issue can be addressed. Because autonomous cars are an emerging technology, it is difficult to know what level of safety can be achieved. Hence, the chapter asks a prior question: How should we think about safety and standards for autonomous cars?

AUTONOMOUS CARS AS AN EMERGING TECHNOLOGY

In discussing safety in autonomous cars, it is important to understand the concept of emerging technologies. Emerging technologies are technologies that are acknowledged to be in the early stages of development. In this respect they are technologies that don't yet exist, at least not in the mature and stabilized form they are expected to achieve. Those involved in their development have visions of what is yet to come. Developers see themselves in a race to get the best, first version into the market so that consumers and the public become familiar with their brand. Consumers are often willing to buy immature versions knowing full well that better models are likely to appear in the future but they want to be first adopters. In addition to autonomous cars, examples of emerging technologies include virtual reality devices, artificial intelligence, brain-computer interfaces, and humanoid robots. Early versions of these technologies are available now, though consumers recognize that these are crude models by comparison with what will be developed in the future.

Because emerging technologies are yet to be fully developed, there is uncertainty about what they will end up looking like in their future, stabilized forms. In the case of autonomous cars, we don't know precisely how the cars will work. That depends on how the technical and social challenges that engineers are working on now are eventually met. We don't know how the cars themselves will work as well as what sort of system they will be part of. Will there be a transition period during which conventional and driverless cars are on the roads together? Safety is much harder to achieve in a mixed system because of the difficulty that self-driving cars have in anticipating the behavior of human drivers. Will there ultimately be *only* self-driving cars on the road? If so, will the cars be able to communicate with one another or will each car react independently to what it senses? Will individuals own cars, or will there be a system (or multiple systems) of cars roaming around so that individuals can order them up as needed much as we now order Uber and Lyft cars?

Emerging technologies present a conundrum. On the one hand, there is no point in investing in the development of something that will never be safe (or cheap or convenient) enough to be accepted and adopted. On the other hand, no one can know for sure whether an idea for a new technology can be developed into something safe (or cheap or convenient) enough *until someone tries*. In the case of autonomous cars, there is enough interest and willingness to invest in the attempt even though there are uncertainties whether the goal can be achieved. Engineers and others must jump in and try to solve the technical challenges to understand whether and how the goal of fully autonomous cars can be achieved.

A NOTE ABOUT TERMINOLOGY

The terms *autonomous, driverless,* and *self-driving* are now being used loosely and often interchangeably even though there is a good deal of variation in the capabilities of current and futuristic models. Current models are not fully autonomous; only certain functions that were previously handled by a human driver have been automated. Technically, these cars and those in the prototype stage are not driverless because drivers are required by law to stay behind the wheel and alert during operation. The term *driverless* is used by some to refer exclusively to cars in the future envisioned to be so fully automated that there will be no steering wheel. To avoid confusion about terms, many in the field rely on a specification of levels of autonomy developed by the Society of Automotive Engineers (SAE). In the United States, the SAE table below is supported by the National Highway and Traffic Safety Association (NHTSA).[1]

> Levels of Automation
> Level 0: The human driver does all the driving.
> Level 1: An advanced driver assistance system (ADAS) on the vehicle can sometimes assist the human driver with either steering or braking/accelerating, but not both simultaneously.

Level 2: An advanced driver assistance system (ADAS) on the vehicle can itself actually control both steering and braking /accelerating simultaneously under some circumstances. The human driver must continue to pay full attention ("monitor the driving environment") at all times and perform the rest of the driving task.

Level 3: An automated driving system (ADS) on the vehicle can itself perform all aspects of the driving task under some circumstances. In those circumstances, the human driver must be ready to take back control at any time when the ADS requests the human driver to do so. In all other circumstances, the human driver performs the driving task.

Level 4: An automated driving system (ADS) on the vehicle can itself perform all driving tasks and monitor the driving environment—essentially, do all the driving—in certain circumstances. The human need not pay attention in those circumstances.

Level 5: An automated driving system (ADS) on the vehicle can do all the driving in all circumstances. The human occupants are just passengers and need never be involved in driving.

Using this system, the vehicles now available for purchase by consumers vary in their level of automation but do not go beyond level 3. In addition to these cars, there are currently a small number of cars on the road with higher levels of automation that are licensed only for experimental purposes. These cars must still have a human behind the wheel at all times. Some say that the term *driverless* should be used only to refer to level 5 cars and *self-driving* only to refer to level 4 cars, but as mentioned earlier, the terms are currently used loosely and often as if they are interchangeable.

THE PROS AND CONS OF AUTONOMOUS CARS

The idea of autonomous cars has grabbed the attention and fascination of engineers, entrepreneurs, the media, and the public. Although there are many reasons for moving to a system of autonomous cars, by far the most compelling argument has to do with safety. Jeffrey Gurney succinctly describes the case as it is commonly made:

Since human driver errors cause most automobile deaths, autonomous vehicles should increase highway safety. The World Health Organization states that around 1.3 million people per year worldwide die from car accidents. In 2010, 32,885 Americans died in car accidents, and over 2.2 million people were injured in a vehicle. Worldwide, Google and other car manufacturers argue that autonomous vehicles will reduce the number of car accident fatalities and injuries resulting from human error because the computer controlling the autonomous vehicle does not get tired, intoxicated, or distracted as does the human driver.[2]

The safety argument simultaneously provides an economic justification. According to a 2012 report, "More than 2.3 million adult drivers and passengers were treated in U.S. emergency rooms in 2009." The report cites the American Automobile Association estimate that car crashes cost Americans $299.5 billion annually. Thus, if autonomous cars reduce accidents, this would enormously reduce the amount of money now being spent for medical treatment.[3]

In addition to safety and the economics of fewer accidents, autonomous cars are also touted on grounds of efficiency and mobility. The idea is that the time people currently spend driving to and from work is not productive, so if individuals are able to work during their commutes, their productivity would be greatly improved. Also, some argue that since traffic jams are one of the things that lengthen commute times and since driverless cars might be controlled in a way that keeps traffic flowing, commute times could be reduced. Moreover, currently a considerable number of people are unable to drive. Driverless cars could mean that the disabled would be able to move about more freely, not having to rely on human drivers.[4]

The arguments in favor of shifting to a system of autonomous cars are, then, very compelling. Yet, on the other side, some argue that research and development are a long way from fully autonomous cars that will be safe enough to be allowed on public roads. Although no one seems to argue that it is inherently impossible to achieve fully autonomous cars that would be safe enough, there is controversy about how long it will take, whether the public will ever accept such cars, and, most important, whether they can ever reach a level of safety that would justify the cost and disruption of shifting to a pure system of autonomous cars.

Those who think that we are a long way from safe self-driving cars cite several daunting technical challenges that are yet to be overcome. For example, although cars currently being tested may operate well in ordinary conditions, they do not yet operate safely in bad weather where there may be dense fog, sudden downpours, and snow accumulation. There are issues about the adequacy of sensors: current sensors cannot detect unusual objects or interpret a traffic policeman's gestures. There are issues about the reliability of machine learning algorithms. Another big problem has to do with the vulnerability of autonomous cars to hacking and spoofing. According to Mary Cummings and Jason Ryan, it is easy to spoof the GPS (global positioning software) systems of autonomous cars, and without proper security, it is feasible for someone to commandeer a car and control its behavior for malicious purposes or just the thrill of it.[5]

In addition to these technical challenges, there are issues involving public acceptance. Some individuals may be reluctant to entrust their lives to autonomous machines. Some already resist the idea of self-driving cars making decisions about who lives and who dies in a collision. Some may be so attached to driving that they will not want to give it up. As well, privacy may be an issue since autonomous cars will likely require cameras inside the car as well as outside, and that will mean surveillance of passengers—if not in real time, then at least in the form of records of passenger behavior.[6] Surveillance will be possible not just through cameras but through the wireless connections that are likely to be required for software updates and other kinds of monitoring.[7] If autonomous cars require an end to privacy while riding, then at least some portion of the public may resist.

THINKING ABOUT SAFETY IN AUTONOMOUS CARS

The potential positives and negatives are, of course, highly speculative at this point since they depend greatly on the concrete ways in which autonomous cars of the future will actually work and what sort of system they will be operate in. In the process of future development, some of the positives and negatives will likely change. For example, the reduction in fatalities currently predicted may shrink as engineers learn about new kinds of failure that can occur in the cars, or the level of trust in self-driving cars may increase as people get accustomed to the idea of not having control. No matter what is learned in the process of development, however, safety will be the central issue. Be it in the technical or social realm, safety will be at the heart of decisions that will determine the success (or failure) of autonomous cars.

Although safety is one of the most important, if not *the* most important, values in engineering, figuring out what is safe enough in the case of autonomous cars is extraordinarily complex. Engineering codes of ethics specify that engineers are to hold the safety of society paramount, and engineering culture and practice make safety a central concern. In fact, safety is so central to engineering that it often goes without saying. A good technology is one that is safe. Good airplanes, appliances, medical devices, buildings, sanitation systems, and so on are considered good in part, at least, because they are safe.

The challenge of figuring out what is safe enough is complex in part because safety is generally not a black or white matter—that is, the question is rarely whether something is safe versus unsafe. Rather the safety issue is whether this technology has an acceptable level of safety or is safer than the alternatives or safe enough on balance given the benefits provided. And these questions are not entirely scientific or engineering matters. Deciding whether something is safe enough or safe on balance is both a technical matter (for example, What are the probabilities that an accident will occur?) and also a matter of values, preferences, and historical and political context (for example, How much risk is acceptable?). So, although engineers are essential to determining the safety of autonomous cars, a range of other actors, including the public and regulatory agencies, are also essential in decision-making about the safety of autonomous cars.

Fatalities involving autonomous cars have already occurred—one in China and several in the United States. These incidents have made the safety issue real, as well as visible to the public. The first accident in the United States occurred in 2016 when a man was killed while his Tesla was in autopilot mode. The car crashed head-on into a tractor-trailer. As the Tesla Team wrote, "Neither Autopilot nor the driver noticed the white side of the tractor-trailer against a brightly lit sky."[8] The second fatality in the United States occurred in March 2018 when a pedestrian crossing the street at night was hit and killed by an Uber vehicle also operating in autopilot mode with a person behind the wheel. The vehicle operator did not see the woman walking with her bicycle. According to one report, the car's radar detected the woman "about six seconds before the crash—first identifying her as an unknown object, then as a vehicle, and then as a bicycle, each time adjusting its expectations for her path of travel."[9] The person behind the wheel had looked away briefly and didn't see the woman in time to react. The woman died later in the hospital.

These incidents are of enormous concern to the public. However, autonomous car advocates argue that the autonomous cars of the future will be much safer than current models. Current models are dangerous in part because drivers are now expected to pay attention and be prepared to take over control in an emergency. In fact, those who sit behind the wheel of cars that are operating autonomously are often referred to as "emergency" or "safety" drivers. This mixed mode of operation is dangerous because emergency drivers are not fully engaged in driving, so when a collision is imminent and the driver is alerted to take over control, the emergency driver may not have time to react quickly enough to avert an accident.[10]

Autonomous car advocates insist that self-driving cars of the future will be so safe that emergency drivers will not be needed. However, there is still the question how we will know when fully autonomous cars are safe enough to operate as such. Who will decide? What criteria will be used to make the safety determination?

Safety and Cost-Benefit Analysis

One of the standard ways that engineers think about safety is in terms of risk. Risk is generally thought of as a combination of the probability of an event occurring times the size of the damage or harm that would result were the event to occur. In this way risk is translated into a number and can be thought of as a cost. The cost can then be put into a cost-benefit analysis.

Cost-benefit analysis has its roots in utilitarianism. Remember that utilitarianism (discussed in chapter 3) is the ethical theory committed to maximizing good consequences. The theory entreats us to identify the bad and good consequences of alternative courses of action and choose the alternative that brings about the most net good. A virtue of this approach is that it is systematic. It gathers different kinds of costs and benefits and combines them into a single measure. In utilitarianism, the unit of analysis is happiness; in engineering, it is generally dollars.

Cost-benefit analysis has an intuitive appeal. Considering the benefits and drawbacks before undertaking any decision or endeavor seems to make good sense. Many people do this informally in their everyday decision-making. Nevertheless, formal uses of cost-benefit analysis have been criticized, and several criticisms are especially applicable to emerging technologies, including autonomous cars.

In the autonomous car debate, cost-benefit analysis is implicit in the idea that the new cars will reduce accidents and thus fatalities and injuries. Such a reduction would be an enormous benefit. Of course, cost-benefit analysis requires that we consider the costs of the reduction as well as the benefits. No one claims that autonomous cars will eliminate accidents; accidents will occur from a variety of factors, such as flaws in mechanical components of the cars, bugs in the software, or simply a deer jumping into the road. So, we know that the 1.3 million lives that are now lost annually from car accidents will not go to zero, but we don't know exactly what the number will be since we don't yet know how the cars will work. In theory, we might consider what size reduction in fatalities—10 percent? 50 percent? 90 percent?—will be worth the costs, but we don't know what the costs will be.

One problem with cost-benefit analysis is that costs are often underestimated and benefits overestimated. This is true, for example, when contractors make bids on construction projects and the projects end up costing much more. Even when the estimates are genuine, the farther into the future that a project goes, the harder it is to accurately calculate costs and benefits. This is especially true for emerging technologies with so many unknowns.

In the case of autonomous cars, some have already begun pointing to some not so obvious effects of autonomous cars, effects that might be put in the cost column. One such effect comes, ironically, from the reduction in accidents. Fewer automobile accidents means fewer organs available for transplant.[11] Others have noted that a system of self-driving cars roaming around waiting to be called would significantly decrease the revenues that cities now obtain from parking meters.[12] These subtle and distant effects are often difficult to foresee and put into monetary terms; though they will affect people, it is not always clear how they can be included in a cost-benefit analysis.

As mentioned, in engineering uses of cost-benefit analysis, costs and benefits are usually put into the single measure of dollar amounts. This is often criticized on grounds that many costs cannot be adequately translated into a dollar amount or any other quantitative measure. The most obvious example is the value of a human life. What dollar amount can or should be put on loss of a life? Even when it comes to injuries, is it enough to consider medical costs as an appropriate measure? As mentioned earlier, in the infamous Pinto case, Ford Motor Company was severely criticized by the public for putting a dollar amount on a human life when it calculated that it would be cheaper for the company to pay off an estimated number of legal suits for loss of a life than to pay for a part that would have made the Pinto less likely to explode in rear-end collisions.

Cost-benefit analyses of autonomous cars will have the same issue. The value of a human life will have to be figured into the calculation of benefits (lives saved) and of costs (lives lost). Other seemingly nonquantifiable values will also have to be dealt with. Freedom and privacy have both been identified as values that are threatened by autonomous cars.[13] Driving an automobile gives people a sense of freedom, and autonomous cars may lead to restrictions on one's freedom to drive. How can we measure the loss of this were people to be compelled to give up driving? As well, if individuals riding in autonomous cars are compelled to give up their privacy, how can that loss be measured? Freedom and privacy are important to democracy, so threats to freedom and privacy are also threats to democracy. Putting a dollar amount on such threats seems impossible.

Yet another problem associated with cost-benefit analysis harks back to an issue with utilitarian theory. Remember that utilitarianism has a problem dealing with distributive justice. The focus in cost-benefit analysis is on total costs and benefits, but costs and benefits fall differently on different people or groups of people. Often certain groups of people receive the benefits of a new technology while other groups bear the costs. Imagine, for example, how a transition to autonomous cars could lead to situations in which those who cannot afford to buy the new driverless cars are at greater risk of accidents than those who can afford to opt into the new system. Or imagine that those who continue to drive are banned from certain areas because self-driving cars are safer when there are no human-driven cars on the road. These entirely plausible scenarios raise serious questions of fairness that simple cost-benefit analysis cannot address.

Uncertainty about the costs and benefits of autonomous cars comes from not knowing both what the cars will be capable of in the future and what kind of system they will operate in. The cars' safety depends both on the cars and the system in which they operate. Will both human-driven and driverless cars operate on the same roads? Will the system consist of roaming cars? A system of cars that communicate with one another and coordinate their behavior to minimize harm? A system of cars in which owners will chose how the cars are programmed to behave?

So, cost-benefit analysis has limited usefulness when it comes to autonomous cars and other emerging technologies. In thinking about safety, cost-benefit analysis should be used cautiously if not skeptically. Part of the difficulty is in the nature of translating costs and benefits into dollar amounts, and part of it comes from the uncertainties that any emerging technology poses. This takes us back to the conundrum of emerging technologies mentioned earlier. We cannot know in advance whether an emerging technology is worth the investment, and yet we will never find out unless we go forward.

Safety and Risk in Sociotechnical Systems

Safety in autonomous cars is generally focused on the cars themselves and this too is a problematic way of thinking. In order to fully understand the safety and risks, it is essential to think of autonomous cars as systems themselves and as systems embedded in broader sociotechnical systems.[14] We tend to think and speak about cars (as well as computers, cellphones, refrigerators, and so on) as if they are merely physical objects—chunks of metal and plastic—and we think of their functionality as coming from the way the material components interact. A good deal of attention is now focused on the sensors, cameras, LIDAR (Laser Illuminating Detection and Ranging) systems, GPS systems, and processors of autonomous cars. These parts will all have to work together, and in this respect autonomous cars are not just a set of inert parts but systems, technological systems. The parts must work together for the car to work, and they have to work together reliably for the cars to be safe.

Cars are not just technological systems; they are sociotechnical systems. For a moment, consider human-driven cars. What makes your car work is not just the engine, ignition switch, tires, and pistons, and it is not just how these parts interact with one another. Your car won't do anything unless you get inside, turn the ignition switch, press pedals, and turn the steering wheel. Your car works through a combination of human and machine behavior. It is a sociotechnical system.

Staying with current cars, you and your car wouldn't get very far, and especially not safely or easily, if it were not for roads, stoplights, gasoline stations, signs, and so on. The roads are designed so that your car fits on them and operates smoothly. Many roads are built so that your car has traction when you press on the brakes, thereby contributing to safety. Gasoline stations allow you to get fuel for your car; stoplights help ensure that you don't easily collide with other cars or pedestrians; and so on. This means that your car works as it does not just because of the car itself (the physical object) but also because of the larger system in which it operates. We sometimes refer to the larger system as infrastructure—your car is part of a transportation infrastructure. Importantly, that infrastructure is not just technological but sociotechnical. The infrastructure consists of things and people, for example, other drivers, pedestrians, gasoline station owners, road repair crews, and traffic enforcers. Your car is also sociotechnical in the sense that people were involved in designing, manufacturing, and marketing it, and people are involved in maintaining your car and the infrastructure in which it operates. So, your car is itself a sociotechnical system as well as part of a larger sociotechnical system.

Now imagine a shift from current cars to fully autonomous (self-driving) cars and suppose that you own one. The shift eliminates you as the driver, but it does not mean that you are entirely out of the system. Your ownership of the car makes it a sociotechnical system, for it means that you are part of the car's operation—not driving but deciding when it operates, where it is parked, how it is insured, when it needs to be taken in for repairs, and so on. And, of course, your car will still be part of the infrastructure that makes the car useful to you—roads, stoplights, signs, gasoline stations, parking areas, auto repair services, and so on.

Even if we suppose that self-driving cars would no longer be privately owned, the people who own or manage the fleets of self-driving cars that circulate would still decide when and where cars are on the road, when the cars need repair, how they are insured, and so on, and people will still be involved in maintaining the communication systems by which cars are ordered up, as well as the systems for providing fuel, regulating traffic, insuring the vehicles, keeping the streets clean, and so on.

A few years back, a cell phone company in the United States ran an advertisement depicting a horde of people huddled around a single person with a cell phone, and the horde followed the person around wherever he went. The horde was meant to illustrate that when a person purchased phone service from this company, the person got not just a phone but an entire support team. Although intended for a different purpose, the image is a striking representation of the fact that it takes many people to make a cell phone work, not just customer support but people to manage computers, cell towers, electric grids, billing processes, and so on. All these people are necessary for the cell phone system to work. The same goes for autonomous cars, regardless of the level of automation.

Recognizing that autonomous cars are components in socio-technical systems is critical to thinking about safety because safety has as much to do with the humans who operate in the system as with the machine parts of the system. The use of the term *autonomous* is unfortunate because it suggests that humans will be out of the loop. However, nothing could be farther from the truth. Even if humans don't do the driving, they will not only interact with the cars—as riders, pedestrians, traffic cops, road workers, and pranksters—but continue to be in control of design-ing, testing, manufacturing, and, perhaps most important, setting safety standards and certifying that cars meet those standards.

Understanding autonomous cars as sociotechnical systems has many implications, one of which has received attention. How the public thinks about the cars and especially their safety is crucial to the success of autonomous cars. Public acceptance is and will continue to be a major issue for autonomous cars, and safety is central to acceptance. Public opinion has been influenced by the accidents that have already occurred with autonomous cars. As a result, some have suggested that in order to garner public acceptance, autonomous cars will have to be at least twice as safe as conventionally driven cars.[15] Others have suggested they will have to be four to five times safer.[16]

Another issue related to public acceptance has to do with the meaning that people associate with driving. When we adopt the sociotechnical perspective and recognize that people are part of transportation systems, we are better able to see that cars and driving have important meaning to people. For example, in the United States, automobiles often serve as status symbols and representations of one's self. Learning to drive is often experienced as a sign of growing up, a point at which one acquires more independence from one's parents. For many, cars are a symbol of their identity; a car may represent a lifestyle, as in the case of a sports cars, or one's wealth, as in the case of Jaguars, or one's political ideology, as in the case of hybrids. So, cars are not just functional devices. This is relevant to the autonomous car debate because some have suggested that at least some drivers will resist giving up driving because of what it means to them. This issue would be easily managed in a system that had a mixture of driver and driverless cars, but it may turn out that a mixed system will be the most dangerous system. If that turns out to be the case, then there will be pressure to prohibit human driving.

Adopting the sociotechnical systems perspective is critical to understanding safety because it allows us to see that safety depends on both machine behavior and human behavior. In the case of autonomous cars, taking humans out of the role of driver doesn't mean that human behavior is irrelevant. One of the most important roles that humans will have is in deciding how safety will be measured and when cars are safe enough to be put on public roads without emergency drivers.

Ensuring Safety through Standards and Regulation

As mentioned, the autonomous cars currently available for purchase and on the road are not fully autonomous; they have level of automation 1, 2, or 3. In addition there are cars on the road with more advanced levels of automation that are licensed for testing on public roads and still require a driver behind the wheel at all times. Just how the cars in experimental mode and those of the future will be determined to be safe enough is an open question. The safety standards for autonomous cars are as much an emerging phenomenon as are the cars.

Some automobile manufacturers are relying on miles driven without an accident to demonstrate safety. Nidhi Kalra and Susan Paddock go to some length to calculate the number of miles (and years) needed to demonstrate autonomous vehicle reliability, but in the end they conclude that "autonomous vehicles would have to be driven hundreds of millions of miles and sometimes hundreds of billions of miles to demonstrate their reliability in terms of fatalities and injuries." They suggest that autonomous car developers cannot "drive their way to safety."[17]

In the United States, the NHTSA has issued guidelines for states to use in assessing whether autonomous cars are ready to be put on public roads, but the guidelines are broad and leave it up to each state to perform its own evaluations.[18] In addition to federal guidelines, many states in the United States have passed legislation regarding autonomous vehicles—not all of it related to safety—and several states have granted licenses to auto manufacturing companies to test experimental self-driving cars.

There seems little doubt that standards, testing, regulations, and legislation will play an important role in ensuring the safety of autonomous cars of the future, but for now it is unclear what approach is the best. Cummings and Ryan have suggested that a good deal can be learned by looking at how safety is achieved with other technologies; they focus on safety in the field of aviation.[19]

The challenge of establishing regulations for autonomous cars is especially daunting because the technology is evolving rapidly. In the interim, Cummings suggests that we think in terms of informed consent. She notes that federal regulation mandates that humans involved in an experiment must be informed and asked for their consent before they are put at risk. Cars that are licensed to operate for experimental purposes are exposing people who are near the cars—for example, drivers, pedestrians, and road crews—to unusual risk. As a step in the direction of informed consent, Cummings argues that experimental cars should be marked as such with a bright color so that those in the vicinity know they are near an experimental vehicle.[20]

THE TROLLEY PROBLEM AND OTHER ETHICAL DESIGN ISSUES

Attention has recently focused on ethical dilemmas in the design of accident avoidance systems. The most discussed dilemma is known as the trolley problem, referencing a dilemma that moral philosophers articulated some time ago in relation to alternative ethical principles. This dilemma was discussed in chapter 3 to illustrate the difference between utilitarian and Kantian thinking. In the philosophical dilemma, a trolley is out of control and headed toward five people who are on the trolley track and will surely die if the trolley continues on its path. However, there is a switch that changes the tracks, and if you flip the switch, the train will be diverted to another track. Unfortunately, there is one person on the alternative track. Thus, if you flip the switch, five people will be saved and one will die. The dilemma is whether you should flip the switch. The justification for doing so is utilitarian; although one death is bad, it is less bad than five deaths. The problem is that in flipping the switch, you will be intentionally causing the death of the person on the alternative track. If you do nothing, five people will die, but you won't be the cause of their death. This dilemma illustrates the difference between killing someone and letting someone die as well as the difference between utilitarianism and Kantian theory or any other theory that is based on intentions. The dilemma is whether it is better to hold to the principle that one should never intentionally kill (even if it means that others will die) or minimize harm even if it means intentionally killing.

The trolley problem manifests itself in decisions about the design of accident avoidance systems in autonomous cars. Presumably, these systems will be so intelligent that in situations in which accidents are imminent, the software will be programmed to calculate alternative ways to maneuver and will be able to choose the alternative that minimizes harm. Imagine a car that detects an imminent collision and calculates that it is better to swerve to the right so as to avoid hitting a car in which there are multiple people, including children, even though in swerving to the right, the car will hit a tree, killing one passenger. Presumably, sensors may be sensitive enough to detect the number of passengers and their size, thus able even to identify children.

The trolley problem aside, accident avoidance systems will have to be programmed to behave one way or another in imminent collisions, so autonomous cars will make ethical decisions, or, more accurately, the designers of accident avoidance systems will make ethical decisions. One uncertainty is whether car manufacturers will let consumer markets determine the design. That is, some consumers may not want to buy a car that will sacrifice them in an accident; others might prefer a utilitarian system that maximizes total good even if it sacrifices them. If these decisions are left to the marketplace, that will mean a variety of accident avoidance systems in operation, and that, in turn, could mean that cars would not readily communicate with one another to minimize harm in dangerous situations.

Another interesting implication of the ethical decision-making of accident avoidance systems is the possibility that people will altogether reject the idea of machines (programs) making life-or-death decisions. This stance has been taken by some in relation to autonomous weapons. There is a significant social movement to ban autonomous weapons.[21] Some people may find it morally preferable for individuals to make ad hoc, spur-of-the-moment decisions in imminent collision situations than to intentionally buy cars with systems that are essentially set to make premeditated choices (formalized in computer software) about who lives and who dies.

The design issues in accident avoidance systems pose deep ethical decisions, and this component of autonomous cars will be important to the safety of the cars and ultimately to public acceptance.

CONCLUSION

Will autonomous cars ever be safe enough? This is a question about the future. We can only speculate and make predictions; no one knows the answer now. We do know, however, that the answer depends on meeting both technical and social challenges. The challenges are not just a matter of developing components that operate reliably, to ensure that autonomous cars are safe; the car itself must consist of reliable parts and must be embedded in a reliable sociotechnical system. It will take reliable machines and reliable people to ensure that autonomous cars are safe.

Though some may be disappointed if a system of fully autonomous cars is never implemented, efforts to develop fully autonomous cars have led to innovations that make conventional cars safer. Examples include autonomous parallel parking systems, advanced cruise control, sensors that detect when a car is changing lanes, and automatic stopping in situations in which a collision is imminent. Although it is not the target, if this trend continues, it is possible that the safety of human-driven cars will improve so much that the justification for autonomous cars will become less compelling.

SUGGESTED FURTHER READING

Cummings, M. L., and J. Ryan. "Point of View: Who Is in Charge? The Promises and Pitfalls of Driverless Cars." *TR News* 292 (2014): 34–37.

Maurer, Markus, et al. *Autonomous Driving.* Berlin: Springer, 2016.

National Transportation Safety Board (NTSB). "Preliminary Report, Highway HWY16FH018." https://www.ntsb.gov/investigations/AccidentReports/Pages/HWY16FH018-preliminary.aspx.

Silberg, Gary, et al. "Self-Driving Cars: The Next Revolution" (KPMG and Center of Automotive Research, 2012). http://www.kpmg-institutes.com/institutes/manufacturing-institute/articles/2017/11/self-driving-cars-the-next-revolution.html.

8 IS SOCIAL JUSTICE IN THE SCOPE OF ENGINEERS' SOCIAL RESPONSIBILITIES?

IN 2015, a major environmental crisis in Flint, Michigan, came to public attention. The city's water supply was contaminated with lead, and residents, especially children, were experiencing severe health effects. Residents had complained about the water for several years, and their complaints had been ignored. The city had struggled economically for decades, and its dire financial situation had led the state to take control. To save money, state authorities decided to switch Flint's water supply from Detroit to the Flint River. Because water from the Flint River is highly corrosive, federal law mandated that an anticorrosive agent be added. However, the Michigan Department of Environmental Quality failed to do this when the switchover was made. Lead from water pipes leached into the water, and residents of Flint were exposed to quantities of lead that far exceeded safe limits. Lead is harmful to all but especially to children; it can cause cognitive impairment, behavioral disorders, hearing problems, and delayed puberty.

The Flint situation was, and still is, a disaster. Adding to the horror were claims of environmental injustice: residents expressed their belief that the contamination would not have occurred had Flint been a wealthy, white city. According to the Census Bureau, over 40 percent of Flint's citizens live below the poverty line and over 50 percent are African American. That many of the concerns persistently expressed by citizens of Flint had been ignored for some time by both state and federal officials suggested a disregard for the health and safety of a poor, largely African American community. In an interview focused on the situation, professor of environmental studies David Pellow explained:

> You and I know that this is an unfortunate but classic case of environmental racism, environmental injustice. Again, we have a community of color, of working-class folks whose well-being, health, and welfare are really not the priority, apparently, for the unelected and appointed emergency manager. Saving money should be a priority. We should always be fiscally responsible. But if you do so at the expense of human health, then we really have a problem. Let's face it, this is a world-class lead poisoning epidemic.[1]

For the purposes of this chapter, the situation in Flint is representative of a broad pattern of inequities in the distribution of social and environmental risks. Access to clean water is only one dimension of this. Environmental injustice takes the form of unequal exposure to toxins from waste disposal sites and nuclear power plants to inequities in air quality, uneven risk of flooding, differences in the distribution of green space, variances in vulnerability to the effects of climate change, and more. Unequal distribution of risk is not unique to the United States; it occurs within many countries all over the world in both developing and industrialized nations.

What, you might ask, does this have to do with engineers and engineering? This chapter asks the question whether social justice is in the scope of engineers' social responsibilities. Understanding the connection between engineering and social justice goes a long way toward answering this question.

THE SOCIAL RESPONSIBILITY OF ENGINEERS

So, how exactly is engineering connected to social justice? A simple answer would go something like this: Engineering endeavors have effects on people—individuals, families, communities, nations, and the global world. Importantly, these endeavors often affect different people differently. An example of this was provided in chapter 7, where we imagined that the adoption of autonomous cars might—depending on the system—negatively affect those who couldn't afford the new cars or those who preferred to drive themselves; many would be made safer from the new system of autonomous cars while some were made worse off or forced to give up something they valued. Another more commonplace example is that of civil engineers siting new buildings, bridges, and roads; the engineers know that in choosing a location, some people will be made worse off and others will benefit. Bridges and roads make it easier to reach certain places—stores, gas stations, neighborhoods—and harder to reach others. Thus, the decision about where to site a new structure often has winners and losers. The distribution of negative and positive effects is especially salient when it comes to citing plants that pose health risks for those living nearby because those residents bear the burden in the form of threats to their health while others benefit from the plants' operation. More broadly, design decisions (which is what a citing decision is) often involve choosing to serve or to better serve some rather than others. For example, the design features of devices may make them harder or easier for particular types of consumers. Think of scissors and chairs designed for right-handed people, websites with text only in English, buildings that are not wheelchair accessible, or machines and tools that are comfortable for use only by people of a certain size or strength.

These examples suggest that the work of engineers can directly affect the distribution of society's benefits and burdens. However, this connection is not widely recognized or acknowledged. To be sure, the engineering codes of ethics recognize that engineering endeavors affect people. This is implicit in the paramountcy clause; the declaration that engineers are to hold paramount the health, safety, and well-being of the public acknowledges that the work of engineers affects human health, safety, and well-being. However, the codes do not include social justice in the list of what should be held paramount. One could argue that social justice is part of human well-being, but that seems a stretch.

Instead of thinking of social justice as falling under human well-being, a better strategy is to think of social justice as part of the social responsibilities of engineers and not to equate engineers' social responsibilities with the paramountcy clause. No one denies that engineers have social responsibilities. Yet there is a problem about social justice because there is no clear consensus on what should be included under the rubric of social responsibilities. Questions about the social responsibilities of engineers have been raised throughout the history of the profession, and especially when new technologies have been developed and used in ways that pose human and environmental risks. After World War II, some asked whether Nazi engineers should have refused to build gas chambers for Hitler. Others criticized those who had worked on the development of the atomic bomb, arguing that the power to kill on such a large scale should never have been created. A variety of social woes have been attributed to technology and hence to engineers. Engineers are implicated in concerns about unsustainable practices, poor air and water quality, threats to personal privacy, deskilling, and more.

Of course, acknowledging that engineers and engineering have social responsibility does not mean that they are responsible only for the bad things that happen. Engineers and engineering are also responsible for a whole host of positive contributions to the good of society. See, for example, the National Academy of Engineering's list of Greatest Engineering Achievements of the 20th Century, which includes electrification, radio and television, the automobile, agricultural mechanization, computers, and the Internet.[2]

Only recently has attention been drawn explicitly to engineers' role in social justice.[3] Engineers create technology and provide technological solutions to an array of problems, and technology and technological decision-making are seen as factors that contribute to social inequities. Moreover, technology is thought not just to contribute to prevailing inequities but to create inequality between current and future generations. The criticism here is that technological decision-making of today may be placing an unfair burden on future generations insofar it disregards consequences for which future generations will have to pay, for example, depletion of natural resources and creation of waste products that will be hazardous for thousands of years. Engineers are not seen as the lone cause of these problems, and they are often viewed positively, as mentioned, as those who take on and solve the big problems of the world.

JUSTICE

Before we delve further into the connection between engineering and social justice, it will be helpful to consider briefly different concepts of justice. At the core, justice has to do with fairness, fairness in the distribution of benefits and burdens. Issues of fairness arise in a wide range of contexts—from the seemingly trivial, as when I get a smaller piece of a delicious pie than you, to the most serious, as when a person who is white but has few qualifications is selected for a job while a person of color with more qualifications is turned down. In many respects, fairness goes to the heart of ethics. Most individuals have a deeply rooted sense of what is fair and how they and others ought to be treated.

Still, figuring out what constitutes justice and how it can be achieved are far from simple matters. Political, economic, and philosophical claims about what constitutes a fair distribution among individuals and groups within a nation or globally are controversial. It seems that identifying social injustice is often easier than figuring out what would be just or identifying how to remedy an unfair situation.

The discussion of ethical theories in chapter 3, especially the point of tension between utilitarian and Kantian ethical theories, serves as a starting place for getting a handle on the notion of social justice. Remember that utilitarianism—at least in its simplest form—has a problem insofar as it seems to justify using some individuals for the sake of many others. This suggests that it is okay for a few to experience negative consequences when doing so will bring about broader, socially beneficial consequences (net utility). According to utilitarianism, decisions ought to be made based on what will maximize overall good consequences, and in certain circumstances that means allowing some to suffer for the sake of others.

One might argue that this is precisely what happens in many situations in which there are social and environmental risks: some bear the risks and many others receive the benefits. For example, in building a toxic waste disposal plant or operating a manufacturing plant that spews pollutants into the air, even though there are negative health consequences for those who live near the plant, the benefits for so many other people seem to counterbalance the negative effects. The question is, Who should bear the risks? In which neighborhoods should those plants be located? Remember that utilitarianism seemed, at least initially, to be ill equipped to handle distributive questions. One can see in the case of locating plants that spew toxic pollutants that although the benefits might outweigh the risks, this doesn't mean that the risks should be borne by one class or one racial group. A good utilitarian would want a distribution that would maximize utility and minimize long-term negative consequences.

Sophisticated advocates of utilitarianism might, for example, consider the long-term negative consequences of a world in which burdens consistently fell to some groups while benefits went to other groups. When pressed to justify the distribution of burdens, it might turn out on analysis that it would be better to impose burdens on those who most benefited or those who were better equipped to handle the risks. It might be argued, for example, that environmental risks should be imposed on the wealthiest and most powerful people because doing so would be more likely to lead to a reduction of risks. That is, if the wealthy and powerful were at risk, they would likely use their wealth and power to find (or put pressure on others to find) safer ways to dispose of waste or more environmentally friendly ways to manufacture products.

Unfortunately, such thinking is rarely used. Instead, decisions seem to be made with more simplistic utilitarian justifications that result in creating or exacerbating social inequities. For example, consider a corporation or a large city government trying to decide where to locate a plant that will unavoidably spew pollutants into the surrounding air. If the decision about where to site the plant is made using a short-term, monetary cost-benefit calculation, the plant will likely be in a low-income neighborhood. Real estate values are lower there, and the neighborhood may be more willing to accept the plant because it will bring needed jobs and tax revenues.

This form of utilitarian reasoning results in what has come to be called the "dumping in Dixie" phenomenon.[4] Low-income residents—often members of minority groups—bear a much greater burden of environmental risk than those who live in wealthy neighborhoods.[5] Decisions to site toxic plants in poor neighborhoods take advantage of existing inequities and worsens them. Those who are already disadvantaged become more disadvantaged. This is part of the cycle of poverty: people who are exposed to air- and waterborne toxins are more likely to become ill, have higher medical expenses, miss work, and perform poorly in school. All of this makes it harder for the poor to get out of poverty. The long-term cumulative effect is that the poor and minority groups bear an unequal portion of social and environmental risks.

Remember that utilitarianism was criticized not just because it leads to unequal distribution but because it treats individuals merely as a means to social good. The counter to utilitarian reasoning was to adopt a Kantian principle, the categorical imperative, according to which all persons are valued as ends in themselves and cannot be used merely as a means to social good. Using the categorical imperative to solve distributional issues seems to call for an equal distribution of risks and benefits or, if that is impossible, then a lottery so that risks are distributed randomly (everyone has an equal chance of bearing the burden). However, Kantian theory also points in another direction. According to the categorical imperative, it might be okay for some individuals to bear more risk than others if those individuals had—with full information—chosen to do so. In Kantian ethics, if some individuals choose to bear a higher proportion of social or environmental risk, then that is okay. However, it is okay only if they have been adequately informed about the risks they are taking and if they are not coerced.

That fairness allows individuals to choose to accept higher risks when they are fully informed and not coerced draws our attention to *how* decisions are made about the distribution of risk. Reflecting on the Flint disaster, one might argue that it was not just utilitarian thinking that led to the problem but the complete disregard for the residents and their complaints. The residents' concerns were treated as invalid and unimportant. This dismissal not only caused the problem to persist much longer than it had to but also created mistrust in state and federal authorities, mistrust that later compounded the difficulty of fixing the problem.

Justice involves more than outcomes. The processes by which a distribution is made must be fair. Individuals who are affected by a distribution must be recognized as ends in themselves. Decisions leading to a distributive outcome must acknowledge the capacity of affected individuals to both know their own interests and values and make decisions for themselves.

Acknowledging that justice involves both outcomes and processes is consistent with a distinction made in the economic, philosophical, and political literature on justice, a distinction between *distributive* and *procedural* justice. Distributive justice focuses on how goods—resources, rights, opportunities—are distributed, and procedural justice focuses on how decisions about distributions are made. Distributive theories of justice focus on principles that should be used in determining a distribution. For example, distributive theories may emphasize equality or merit as principles for distribution. Procedural justice focuses on starting conditions for bargaining and who should be involved in the decision making and how. From a real-world perspective, the distinction is somewhat tricky. Procedural justice advocates argue that if we focus only on an actual distribution, we may fail to see that a distribution—even though unequal—was fair because it was arrived at through a fair process. Distributive justice advocates argue, on the other hand, that if a procedure results in what looks like an unfair outcome—as in the case of distributions that fall along racial or gender lines when these characteristics are irrelevant—then the procedure is likely unfair.

So, distributive justice theorists looking at who currently bears the bulk of environmental burdens would likely conclude that present conditions are unfair. On the other hand, procedural justice theorists would have to look at how the decisions that led to the current distribution were made. For example, individuals may have moved to neighborhoods knowing that a toxic plant was there, or communities may have accepted the building of a dangerous plant because they were promised economic benefits—jobs, tax revenues. Of course, these individuals may have accepted the risks because they were in dire circumstances that gave them no good choices. Procedural justice advocates could then raise the question whether the apparent consent of these local individuals was valid. Attention would focus on the fact that these individuals chose while in dire circumstances and would not have chosen to incur the risks were they better situated. So, procedural justice theorists would not declare the current situation unjust merely because of the existing distribution; they would look at the processes that led to the distribution.

Although the debate between these two types of theories of justice cannot be settled here, it is important to note that both distributive and procedural issues come into play in considering social and environmental injustice. Questions have been raised and arguments made suggesting both that the current distribution of social and environmental risks is unfair and that the decision-making processes leading to the current distribution were unfair.

Another notion of justice worthy of mention here is *compensatory* justice. When individuals sue other individuals or entities because a harm was inflicted on them, they seek compensation for the harm done. Although the harm cannot be undone (for example, loss of a loved one, severe illness caused by toxic exposure, or loss of reputation), the plaintiff seeks compensatory justice. Compensatory justice is often implicit or explicit in discussions of social justice because of the long histories of ill treatment of certain groups of people—racial, religious, ethnic, and gender groups. Hence, social and environmental justice has a compensatory valence. Some environmental justice advocates argue that we should undertake actions now that will not just bring about a better balance of risks and benefits but will compensate those who were unfairly environmentally disadvantaged in the past.

Yet another perspective on social justice, one often used in discussions of social and environmental justice, is that of *human rights*. Human rights come into play in social and environmental justice issues as a foundation for criticisms of practices that fail to recognize fundamental human rights. In the human rights tradition, it is taken as a given that human beings are entitled to certain things because they are human beings. For example, the United Nations has issued Millennium Development Goals, described as "basic human rights—the rights of each person on the planet to health, education, shelter, and security."[6] The Millennium Development Goals have eight targets:

1. Eradicate extreme hunger and poverty
2. Achieve universal primary education
3. Promote gender equality and empower women
4. Reduce child mortality
5. Improve maternal health
6. Combat HIV/AIDS, malaria and other diseases
7. Ensure environmental sustainability
8. Develop a global partnership for development

In the human rights tradition, social justice is a matter of achieving these conditions for all, and anything short of this constitutes injustice.

So, justice is not a simple concept, but it is an important concept that focuses attention on patterns of distribution of benefits and risks (burdens). It gives us a way—albeit several ways—of thinking about what we mean when we say that a situation is unjust or unfair. Moreover, the term *social justice* provides a way of expressing a goal or set of goals that have to do with making the world a better place. Donna Riley articulates this idea succinctly when, after reviewing various definitions of social justice, she explains that the "theme that cuts across these definitions is the struggle to end different kinds of oppression, to create economic equality, to uphold human rights or dignity, and to restore right relationships among all people and the environment."[7]

IS SOCIAL JUSTICE IN THE SCOPE OF ENGINEERS' SOCIAL RESPONSIBILITIES?

With this clearer, though more complex, understanding of justice, we can now return to the connection between social justice and engineers' social responsibilities. Earlier we established that the connection was grounded in the fact that the work of engineers has social effects. The connection to social justice derives from the fact that those effects can create or exacerbate social inequities or contribute to equalizing them. The work of engineers affects different people differently and in so doing can affect, negatively or positively, the distribution of benefits and burdens in society.

Technology, Engineering, and Society

Another, perhaps more abstract, way to think about the connection between engineering and social justice is to consider the triadic relation of technology, engineering, and society. *Engineers* contribute to the creation of *technology*. They design, monitor, maintain, assess, work with, and understand technology. Technology, in turn, constitutes and structures *society*. The connection between engineering and technology almost goes without saying. That is, it is commonplace to think of technology as the purview of engineers. The connection between technology and society is less well recognized or understood, though awareness of a connection has increasingly been recognized in the last half century. Oddly, however, the implication of these two connections taken together is often neglected. A common assumption is that engineers are and should be concerned only with technology; their business is solving technical problems, not societal conditions.

The problem with this view is that the decisions engineers make affect social arrangements, social relationships, and social values, and it is much harder to undo technologies when we later learn of their negative social effects. In other words, if engineers ignore the social consequences of their work, all kinds of bad consequences that could have been avoided may result.

Technology affects so many aspects of our lives that it is challenging to think of a domain in which technology plays no role. Think of how the Internet has changed who we interact with, how often, and when. Technology orders social arrangements; some would say that technology governs how we live our lives—think here about clocks, stoplights, sidewalks, and roads. Technology shapes how we understand our place in the world—think here about how telescopes and space travel have changed the way individuals orient themselves on Earth. Technology influences ideas about what it means to be human—consider how such radically different technologies such as artificial intelligence and ultrasound (used to view fetuses) have affected our conceptions of personhood. Once one embraces the idea that technology and society are thoroughly intertwined, the need for a reconceptualization of engineering seems necessary. Engineers aren't just solving technical problems and they aren't just creating artifacts and devices: they are making society.

In 1980, Langdon Winner introduced an idea that has resonated for decades in the field of science and technology studies. The idea is that artifacts (material objects) have politics.[8] In illustrating this idea Winner provided an example of how the material design of bridges had social justice implications. Winner argues that Robert Moses, an influential developer in New York in the 1930s, had designed the bridges of Long Island, New York, to be at a height that would not allow public buses to fit below the underpasses.[9] According to Winner, the height of the bridges constrained bus routes and meant that buses traveling from New York City could not reach the pristine beaches of Long Island. This meant that poor city dwellers, including many African Americans, could not get to those beaches in summer. They didn't own cars, so the only way they could reach the beaches was by public transportation. Winner's argument was that the design of the bridges—their materiality—reinforced the prevailing social hierarchy. It kept poor and often African American people off the beaches.

Winner has been challenged on the details of this account and in particular on the issue of whether Moses consciously intended to reinforce the racial and class divide. Regardless of Moses's intentions, however, the story saliently illustrates how the height of a bridge can influence social order. Material features of the bridge reinforced and solidified a racial and wealth hierarchy. No one was formally prohibited from using the beaches, but access was made much harder for those without cars.

Winner's argument seems to dispel the claim that technology is value neutral. For many decades, the idea that technology was value neutral was much debated by those who were thinking about the social implications of technology. Those who held that technology is neutral claimed that artifacts acquire value only when used by humans. Perhaps the most familiar version of this argument is the claim that guns don't kill people, people kill people. Winner's thesis generated a conversation that has led to several ideas about how material objects and complex technological systems affect human behavior. The newer theories suggest that material objects constrain and enable various behaviors. Thus, many now believe that the guns versus people debate is a false dilemma. People and technological objects such as guns are so intertwined that they can't be disentangled.

In discussing a course on "Engineering, The Environment, and Society" that he teaches at the University of California, Berkeley, civil and environmental engineer Khalid Kadir explains: "Technical experts draw a box around a technical problem. We call it a control volume. . . . We have inputs and outputs and we deal with what's inside the box. So when we draw the box around water in the Central Valley that contains nitrate, we don't look at undocumented laborers, we don't look at substandard housing, we don't look at that larger picture because that's not what our training tells us to do. We are there to deal with nitrate in drinking water. I started unpacking that in my own work and started asking about the bigger picture. . . . How do we make engineering part of larger solutions and not just technology pieces that we throw in and leave?"[10]

Many engineers today are aware of some of the social effects of their decisions and take these into account in their decision making. Perhaps the most obvious case, mentioned earlier, is that of civil engineers who decide (or recommend) where to build roads and bridges. They know that their decisions will be beneficial to some and less beneficial or even harmful to others. But civil engineers aren't the only engineers who recognize the social implications of their work. Biomedical engineers are aware that the medicines and devices they design may be well suited for certain types of bodies and not for others; for example, one gender or age group or ethnic group may be helped while others are not.

Similarly, computer scientists are aware that websites can be designed in ways that make it easier or harder for users with older equipment to access sites or for users who don't understand English. These examples show that engineers are aware of how their work affects people even though the engineers often see themselves as focused exclusively on technology and technical matters.

Once the connections between the work of engineers and technology and society are made evident, the fact that the work of engineers can have implications for social justice also becomes undeniable. Nevertheless, there are some who will argue, and rightly so, that it's not so simple. Yes, engineers have effects on social justice, but most engineers are not able to do anything about it; hence, they shouldn't be held or considered responsible for it.

Two Positions

At this point, it does not seem plausible to argue that the work of engineers has no connection to social justice. However, it also seems unfair to hold engineers responsible for something they cannot control. Many, if not most, engineers have little or no authority over the decisions that have the greatest effects on social justice. Engineers generally work in organizational contexts in which they may have little say about what projects they work on, for whom, and at what price. They work for employers and clients who make the big and overarching decisions about what will get made, what general features it will have, who the customers will be, and so on. So, the work life of many engineers is restricted to making narrow technical decisions. Those who are higher up in organizational hierarchies—owners, managers, and executives— are the people who make the decisions that affect social equity.

Indeed, in addition to owners, managers, and executives, many other actors are involved in decisions that affect social justice. This includes those who shape and regulate the markets in which engineering products and services are distributed (policy makers), those who make trade agreements (politicians), those who engage in corrupt practices that siphon off resources from engineering endeavors (criminals), and many others. Another way of saying this is simply to say that powerful individuals (who may or may not be engineers) and economic and political systems are the more potent elements affecting the distribution of risks and benefits from engineering endeavors.

That engineers often work in organizations in which they have little power to do anything about the social justice implications of their work is undeniable. However, to infer from this that engineers have no responsibility for social justice is an overstatement. The argument takes us back to chapter 4, in which the question whether engineers should think of themselves as guns for hire was raised. To say that engineers should leave the social implications of their work entirely up to their employers is comparable to saying that they should allow their expertise to be deployed for any purposes for which someone is willing to pay. As suggested in chapter 4, this is problematic because it suggests that engineers should cease to be moral agents when they go to work. It effectively reduces engineers to machines since machines have no sense of right or wrong, good and bad. Such a position seems contrary to the very notion of morality, and it is certainly contrary to the values of the engineering profession as expressed by professional societies in their codes of ethics.

The alternative is to embrace the idea that engineers, individually and collectively, bear some (though not all) responsibility for social justice and then focus on figuring out what engineers might be able to do to address the matter. This position can acknowledge that individual engineers may not have a good deal of power in the organizations within which they work, but they still have a role to play in considering the effects of their work. According to this position, engineers have a responsibility to do whatever they can to prevent social injustice consequences and promote social justice consequences from their projects.

KEEPING SOCIAL CONSEQUENCES IN SIGHT

So, what can engineers do, given their limited decision-making power in the organizations within which they work? First and most important, engineers should acknowledge (to themselves and others) that their work has social justice implications. Although they generally see themselves as problem solvers, engineers generally don't see themselves as *social* problem solvers. They may not see that technical problems *are* social and human problems. Even something as technical as writing code for software is directed at fulfilling a human need or desire—the software is part of a device that allows users to better do whatever they seek to do. Projects that seem utterly disconnected to people can always be traced back to a human interest or purpose. For example, designing corrosion reduction methods for turbines may seem to be a purely technological endeavor, but it is a human desire to increase efficiency and endurance when performing a task that humans want done that ultimately motivates the effort.

Perhaps the most salient way to see that engineers are solving social problems while solving technical problems is to consider the list of Grand Challenges put forward by the National Academy of Engineering. In 2013, the academy convened a committee to develop ideas about the greatest challenges and opportunities for engineering in the twenty-first century. They came up with fourteen:

1. Make solar energy affordable.
2. Provide energy from fusion.
3. Develop carbon sequestration methods.
4. Manage the nitrogen cycle.
5. Provide access to clean water.
6. Restore and improve urban infrastructure.
7. Advance health informatics.
8. Engineer better medicines.
9. Reverse-engineer the brain.
10. Prevent nuclear terror.
11. Secure cyberspace.
12. Enhance virtual reality.
13. Advance personalized learning.
14. Engineer the tools for scientific discovery.

Although each of these challenges are specified in a way that makes them solvable by means of engineering, each responds to a human need or human problem and each is targeted to make for better human lives.

So, one strategy to facilitate engineers in not creating or exacerbating social inequities and promoting social justice is to embrace the idea that engineers are building, not just things, but also society. The products, services, systems, and endeavors to which engineers contribute affect society; they solve social problems and configure or reconfigure the world in which people live. The work of engineers shapes social arrangements, social relationships, and conditions of human living.

ENGINEERING, DESIGN, AND SOCIAL JUSTICE

In harmony with acknowledging the social and sometimes social justice implications of their individual work, engineers can join forces with other engineers both within and outside their jobs. Although the analysis of this chapter has focused on the work of individual engineers, it is important to note that the connection between engineering and social justice also points to an important role for collectivities of engineers taking responsibility for and using their expertise to promote social justice. Whether through formal engineering societies or informal organizations of engineers, there is much that engineers can do collectively. A good example of how engineers have collectively and individually taken initiative in volunteering their expertise outside the workplace is the organization Engineers without Borders (EWB). EWB involves engineers and engineering students in projects that "empower communities to meet their basic human needs and equip leaders to solve the world's most pressing challenges."[11]

EWB is just one organization representing the idea that engineers can address issues of social justice outside their ordinary work life. Engineers can contribute their expertise to local, national, and international projects, providing services to those who might not otherwise have access to engineering expertise.

One important but often ignored way that individual engineers can affect social justice is in their choice of where to work and what projects to work on. Too little attention is paid to the ethical implications of an engineer's choice of employer and/or choice of research area. Yet whom one works for as well as whom one works with can make all the difference in the contributions one ultimately makes in one's career. Working for companies and clients that are committed to solving social problems and care about social justice means that one's work will have a different impact than working for companies that care only about the bottom line.

Engineers who are involved in design have a special opportunity to influence social justice. When engineers design products, be they medical devices, software, chemical processes, cars, or airplanes, awareness of the broad social implications of this work can make a difference. In designing products or systems or solutions, engineers generally have some sense of who will use what they are creating and how. When it comes to products to be used by consumers, for example, engineers can be careful about whom they have in mind as the users. If they think about users only as people like themselves, they may, for example, design only for white, male, middle-class Americans. Simply being aware that lots of people live differently from oneself can go a long way in affecting the products that engineers develop. Examples of this are easy to find. Only after public awareness and legislation penetrated the work of engineers was attention paid to the needs of the disabled. For decades, engineers designed buildings with entrances at the top of steps and sidewalks too narrow for wheelchairs. They weren't thinking of people with disabilities.

Another case that illustrates the point in terms of gender is a well-known case involving the design of an airplane cockpit. In this case engineers were told to design the cockpit so that it would accommodate 90 percent of pilots (excluding the tallest and shortest 10 percent). Without much thought, they used standard charts for human heights and found out only later that the charts they used were for the heights of men. When women applied to be pilots, many were too short to fit safely in the cockpit, and this was used to disqualify them from becoming pilots.[12] The point is that simply thinking carefully about who is going to use a product and having in mind the diversity of human beings and conditions of life can make a difference in the design of technologies and who ultimately can use them. This can contribute to social justice.

Women and Minorities in Engineering

Another way that engineers (individually and collectively) can and should promote social justice is closer to home. The social justice lens can be used to view the engineering profession itself. Today relatively few women and minorities are engineers or studying to become engineers. It appears as if women and minorities are being denied the opportunity to become engineers, an opportunity to earn a decent living while doing challenging and interesting work. The issue is complicated because evidence suggests that the skew in representation is not just that women and minorities are explicitly being denied entrance to the profession; rather, many choose not to go into engineering even if they have the appropriate qualifications. This situation has received a good deal of public attention, in part because there is projected to be a shortage of engineers in the future. National, international, and local efforts are being made to bring more women and minorities into engineering (and other related fields). These efforts have had modest success, though in some subfields more than others.

Concerns about why there are so few women and minorities in the field hint at looking back and blaming practices that may have explicitly or implicitly discriminated against these groups. However, the backward-looking approach is relevant here only insofar as it points to practices that are unnecessary and can be changed to bring more minorities and women into engineering. There may be or may have been bias in engineering school admission procedures or corporate or academic hiring and promotion processes. These must be examined, as should aspects of engineering culture that make it uncomfortable for women and minorities to work in the field.

The makeup of the engineering profession has broader implications for social justice. Not having women and minorities in engineering may well affect which problems get addressed and who is served by engineering. Some argue that the needs of women and minorities are less likely to be addressed when they make up such a small proportion of the profession. The idea here is that because women and minorities have life experiences that are unique, they are likely to choose different problems to work on or take approaches to problem solving that are more likely to serve people like them. Women and minority engineers are more likely to have people like them in mind when they design and problem solve. The same can be said for those who come from different cultures and socioeconomic groups. It may well be that the interests of these groups are more likely to be addressed if there are engineers who have had the experience of being poor, coming from less developed parts of the world, being a minority, and so on.

The small number of women and minorities in engineering gives us yet another way in which engineers can address social justice—that is, by reflecting on their own field and working environment. Individual engineers can make a difference here simply by ensuring that they are contributing to making a comfortable climate for a diversity of engineers.

In addition to these ways of individually addressing issues of social justice at work, engineers can engage in activities outside work. Engineers can volunteer their services to community groups or work on professional society initiatives that tackle the needs of those who might not otherwise have the benefit of engineering expertise.

CONCLUSION

Is social justice in the scope of engineers' social responsibility? This chapter has argued that it is. Although the culture of engineering often suggests that engineers are narrowly focused on machines and devices, the work of engineers makes society what it is. The work of engineers contributes to the character of society, including the distribution of benefits and burdens. The extent of an individual engineer's power to control the social effects of his or her work varies a good deal depending on what position the engineer holds and what project the engineer works on. Limitations on what engineers can do are not, however, a justification for dismissing the challenge to use whatever power they have to address issues of social justice.

We began this chapter with the accusation of environmental injustice in Flint, Michigan. Although the lead problem persists, with many people still lacking clean water in their homes, some progress has been made. As it happens, an engineer played an important role in identifying and documenting the problem. University professor Marc Edwards worked with residents to collect and analyze water samples to demonstrate to resistant authorities that there was a problem and to document its severity. A few years earlier, Edwards played a similar role in documenting elevated lead levels in the municipal water supply of Washington, DC. Asked about his role in Flint, Edwards said, "I didn't get in this field to stand by and let science be used to poison little kids." He continued, "I can't live in a world where that happens. I won't live in that world."[13]

SUGGESTED FURTHER READING

Mohai, Paul, David Pellow, and J. Timmons Roberts. "Environmental Justice." *Annual Review of Environment and Resources* 34 (2009): 405–30.

Nieusma, Dean, and Donna Riley. "Designs on Development: Engineering, Globalization, and Social Justice." *Engineering Studies* 2, no. 1 (2010): 29–59.

Riley, Donna. *Engineering and Social Justice.* San Rafael, CA: Morgan and Claypool, 2008.

CONCLUSION

THE subtitle of this book is *Contemporary and Enduring Debates*. You might think that the foundational issues described in part I are the enduring debates and that the debates presented in part III are contemporary ones. That would be a mistake. All eight chapters take up enduring issues, though some do so in the guise of a contemporary controversy. For example, although the focus on an emerging technology, autonomous cars, in chapter 7 may seem a thoroughly and uniquely contemporary topic, safety and risk are enduring (and central) issues in engineering. Debates about the safety of technologies and the acceptable level of risk have persisted in engineering from its beginnings, though the nature of particular safety issues has changed as new technologies have emerged. Think, for example, of airplane travel, nuclear power, and genetically modified food. Even chapters 4 and 5, which are about employment relationships, raise general questions that endure: What do engineers owe their employers and clients? How far should engineers be expected to go to protect the public? These questions have had to be asked anew as norms and expectations for employment relationships, the structure and scale of organizations, legislation, and engineering professional societies have evolved over time.

The eight chapters of this book do not cover every ethical issue that can arise in engineering, nor do they include all the issues typically considered part of the field of engineering ethics. The topics selected here are representative of a much larger panoply of ethical issues that face engineers, the engineering profession, and any society configured with technology. They were selected for reasons mentioned in the Introduction. Among these, the selection was targeted to include micro-ethics (ethical issues faced by individual engineers) and macro-ethics (ethical issues that face the engineering profession and issues of public concern and in need of public policy). The topics were also chosen to include issues relevant to engineers from a variety of subfields, including civil, mechanical, electrical, and biomedical engineering.

The analysis presented in each chapter is meant to model analytical strategies that can be used to address other issues as well as to provide a basis for anticipating and responding to issues that will take shape in the short- and long-term future. One such analytical strategy is the debate format. This strategy has been used to ensure that issues are viewed from diverse perspectives; it is a strategy that is helpful in addressing any ethical issue. The debate format has also been used to ensure that none of the tenets of engineering ethics could be taken as dead dogma. Ethics is rarely a matter of receiving simple rules and blindly following them. Doing the right thing is often quite complex; it involves incorporating facts as well as norms and values and using judgment. Importantly, doing the right thing in engineering involves understanding complex technical and social circumstances.

Readers may be surprised or even disappointed that the conclusions drawn at the end of each chapter do not always consist of a definitive position on the chapter's central question. My goal in writing these chapters was not to convince the reader that a particular position is the only or best possible position. Rather, the idea was to sort out wheat from chaff and leave the reader convinced, not only that certain ways of thinking about an issue are implausible or unacceptable, but also that arriving at a definitive answer might require more analysis and might depend on specific details about particular situations. For example, deciding whether whistleblowing engineers are all heroes or traitors cannot be decided in general: each case must be taken on its own.

Considering all eight chapters together, three claims might be thought of as the underlying lessons of the book. They are lessons that I hope engineers and others will embrace because they make the ethical issues in engineering more obvious. First, *Engineering is a social endeavor.* Nothing is achieved by engineering without human actions, social relationships, organizations, law, and social values. Thinking about engineering as merely a technological endeavor is not only a false picture but also hides the centrality of ethics to the field. Ethical engineering involves managing multiple social relationships, including relationships with the public and those affected by engineers' work, as well as relationships with employers, clients, contractors, regulators, other engineers, and many more.

Second, *Engineers do not just make things; they build society.* Technology neither comes out of a social vacuum nor goes into a social vacuum. Technology constitutes the world that people live in and in so doing affects people, social arrangements, and social values in powerful ways. This is most saliently evident when new technologies are introduced, as when our social relationships are reconfigured around cell phones, but it is also true of older technologies, for they too constrain and enable what we do.

The third lesson follows from the first two. *Engineering is inherently an ethical enterprise.* Because engineering is a social endeavor and because engineers are building the society in which people live their lives, the behavior of engineers falls within the domain of ethics. To put this succinctly: because of its social nature, engineering should be viewed through the lens of ethical analysis.

NOTES

INTRODUCTION

1. See Terry S. Reynolds, ed., *The Engineer in America: A Historical Anthology from Technology and Culture* (Chicago: University of Chicago Press, 1991).

2. Philip L. Alger, N. A. Christensen, and Sterling P. Olmsted, *Ethical Problems in Engineering* (New York: John Wiley and Sons, 1965).

3. See Carl Mitcham, "A Historico-Ethical Perspective on Engineering Education: From Use and Convenience to Policy Engagement," *Engineering Studies* 1, no. 1 (2009): 35–53.

4. See John Stuart Mill, *On Liberty* (London, 1859), chapter 2.

1. CAN ENGINEERING ETHICS BE TAUGHT?

An earlier version of this chapter was published as D. G. Johnson, "Can Engineering Ethics Be Taught?," The Bridge 47, no. 1 (March 3, 2017), available at https://www.nae.edu/19582/Bridge/168631/168649.aspx.

1. Peter Worthington, "Comment," *Toronto Sun*, June 6, 2010.

2. Karl D. Stephan, "Can Engineering Ethics Be Taught?," *IEEE Technology and Society Magazine* 23, no. 1 (2004): 5.

3. Originally ABET was an acronym for Accreditation Board for Engineering and Technology. In 2005, the organization changed its official name to ABET.

4. Jonathan Haidt, "The Emotional Dog and Its Rational Tail: A Social Intuitionist Approach to Moral Judgment," *Psychological Review* 108, no. 4 (2001): 817.

5. David A. Pizarro and Paul Bloom, "The Intelligence of the Moral Intuitions: Comment on Haidt (2001)," *Psychological Review* 110, no. 1 (2003): 195.

6. This case and the Board of Ethical Review commentary are in the National Society of Professional Engineers' repository of cases, www.nspe.org/resources/ethics/ethics-resources/board-of-ethical-review-cases. Reprinted by Permission of the National Society of Professional Engineers (NSPE), www.nspe.org.

7. S. Eriksson, G. Helgesson and A. T. Höglund, "Being, Doing, and Knowing: Developing Ethical Competence in Health Care," *Journal of Academic Ethics* 5, no. 2–4 (2007): 207–16.

8. Charles J. Abaté, "Should Engineering Ethics Be Taught?," *Science and Engineering Ethics* 17, no. 3 (2011): 583–96.

9. The situation described here is based on Case 23 in Charles E. Harris, Michael S. Pritchard, and Michael J. Rabins, *Engineering Ethics: Cases and Concepts*, 2nd ed. (Belmont, CA: Wadsworth, 2000), 316–17.

10. Abaté, "Should Engineering Ethics Be Taught?," 591.

11. Courses on engineering ethics began to be taught as a formal subject matter in undergraduate education in the 1980s.

12. Michael S. Pritchard, "Professional Responsibility: Focusing on the Exemplary," *Science and Engineering Ethics* 4, no. 2 (1998): 215–33; Chuck Huff and Laura Barnard, "Good Computing: Life Stories of Moral Exemplars in the Computing Profession," *IEEE Technology and Society Magazine* 28, no. 3 (2009): 47–54.

13. Pritchard, "Professional Responsibility," 221.

14. Ibid.; Huff and Barnard, "Good Computing"; Chandrakant B. Madhav, "Phronesis in Defense Engineering: A Case Study in the Heroic Actions of Two Defence Engineers as they Navigate Their Careers" (Ed.D. diss, University of St. Thomas, Minnesota, 2014).

2. DO ENGINEERS NEED CODES OF ETHICS?

1. The complete NSPE Code of Ethics is available online at: https://www.nspe.org/resources/ethics/code-ethics.

2. The ASCE Code of Ethics is available online at: http://www.asce.org/code-of-ethics/.

3. The IEEE Code of Ethics is available online at: https://www.ieee.org/about/corporate/governance/p7-8.html.

4. The ASME Code of Ethics is available online at: https://www.asme.org/about-asme/advocacy-government-relations/ethics-in-engineering.

5. Elizabeth H. Gorman and Rebecca L. Sandefur, "Golden Age, Quiescence, and Revival: How the Sociology of Professions Became the Study of Knowledge-Based Work," *Work and Occupations* 38, no. 3 (2011): 275–302.

6. This section focuses on practices in the United States, though many countries of the world organize engineering in a similar way.

7. "What Is a PE?," NSPE, http://www.nspe.org/resources/licensure/what-pe.

8. Ibid. for a full list of the rights and responsibilities of professional engineers.

9. "Frequently Asked Questions about Engineering," NSPE, http://www.nspe.org/resources/media/resources/frequently-asked-questions-about-engineering.

10. Sarah K. A. Pfatteicher, "Depending on Character: ASCE Shapes Its First Code of Ethics," *Journal of Professional Issues in Engineering Education and Practice* 129, no. 1 (2003): 21.

11. Mike W. Martin and Roland Schinzinger, *Introduction to Engineering Ethics*, 2nd ed. (New York: McGraw-Hill Higher Education, 2010), 41.

12. Charles E. Harris Jr., "Internationalizing Professional Codes in Engineering," *Science and Engineering Ethics* 10, no. 3 (2004): 503–21; Jathan Sadowski, "Leaning on the Ethical Crutch: A Critique of Codes of Ethics," *IEEE Technology and Society Magazine* 33, no. 4 (2014): 44–72.

13. Diane Michelfelder and Sharon A. Jones, "Sustaining Engineering Codes of Ethics for the Twenty-First Century," *Science and Engineering Ethics* 19, no. 1 (2013): 237–58.

14. Harris, "Internationalizing Professional Codes in Engineering."

3. HOW SHOULD ENGINEERS THINK ABOUT ETHICS?

1. Charles E. Harris Jr., "The Good Engineer: Giving Virtue Its Due in Engineering Ethics," *Science and Engineering Ethics* 14, no. 2 (2008): 153–64.

2. Caroline Whitbeck, *Ethics in Engineering Practice and Research* (Cambridge: Cambridge University Press, 2011).

4. SHOULD ENGINEERS SEE THEMSELVES AS GUNS FOR HIRE?

1. NSPE Case No. 13-11, www.nspe.org/resources/ethics/ethics-resources/board-of-ethical-review-cases. Reprinted by Permission of the National Society of Professional Engineers (NSPE), www.nspe.org.

2. NSPE Case No. 15-8. Reprinted by Permission of the National Society of Professional Engineers (NSPE), www.nspe.org.

3. NSPE Case No. 09-7. Reprinted by Permission of the National Society of Professional Engineers (NSPE), www.nspe.org.

4. Transparency International, https://www.transparency.org/what-is-corruption#define.

5. See Nicholas Ambraseys and Roger Bilham, "Corruption Kills," *Nature* 469 (2011): 153–55.

6. See https://www.constructiondive.com/news/former-mta-construc
tion-manager-fined-sentenced-to-prison-on-bribery-charg/517896/.

7. Brendan Pierson, "N.Y. Developer Pleads Guilty ahead of 'Buffalo
Billion' Corruption Trial," *Reuters*, May 18, 2018, https://www.reuters.com
/article/us-new-york-corruption/n-y-developer-pleads-guilty-ahead-of-buffalo
-billion-corruption-trial-idUSKCN1IJ2TQ.

8. See https://blog.capterra.com/construction-fraud-stories/; and
Stephanie Clifford, "Construction Company Admits to Defrauding New
York Clients," *New York Times*, May 20, 2015.

9. See Daniel Gallas, "Brazil's Odebrecht Corruption Scandal," *BBC
News*, March 7, 2017, https://www.bbc.com/news/business-39194395.

10. William P. Henry, "Addressing Corruption in the Global Engineer-
ing/Construction Industry," *The Bridge* 47, no. 1 (Spring 2017): 52–58,
https://www.nae.edu/19582/Bridge/168631.aspx.

5. ARE WHISTLEBLOWING ENGINEERS HEROES OR TRAITORS?

1. Jean Kumagai, "The Whistle-Blower's Dilemma," *IEEE Spectrum*,
April 1, 2004, http://spectrum.ieee.org/at-work/tech-careers/the-whistle
blowers-dilemma.

2. Vitrification is the process of transforming materials into a glass
form.

3. Ralph Vartabedian, "Nuclear Whistle-Blower Backed by Watchdog
Agency," *Los Angeles Times*, July 6, 2011.

4. This account is based on Walter Tomasaitis's testimony before
the US Senate Homeland Security Subcommittee on Financial and
Contracting Oversight and on Annette Cary, "Hanford Whistleblower
Tamosaitis Loses His Job," *Tri-City Herald* (Kennewick, WA), October 4,
2013. For Tomasaitis's testimony, see "Whistleblower Retaliation at the
Hanford Nuclear Site," March 11, 2014, Senate Hearing 113-370, https://
www.gpo.gov/fdsys/pkg/CHRG-113shrg88280/html/CHRG-113shrg88280.
htm.

5. Siddhartha Dasgupta and Ankit Kesharwani, "Whistleblowing:
A Survey of Literature," *IUP Journal of Corporate Governance* 9, no. 4
(2010): 57.

6. Mike W. Martin and Roland Schinzinger, *Introduction to
Engineering Ethics*, 2nd ed. (New York: McGraw-Hill Higher Education,
2010), 161.

7. Juan M. Elegido, "Does It Make Sense to Be a Loyal Employee?,"
Journal of Business Ethics 116, no. 3 (2013): 495.

8. Richard De George, "Ethical Responsibilities of Engineers in
Large Organizations: The Pinto Case," *Business and Professional Ethics
Journal* 1, no. 1 (1981): 6.

9. Ibid.

10. Editorial Board, "Edward Snowden, Whistle-Blower," *New York Times*, January 1, 2014.

11. US House of Representatives, *Review of the Unauthorized Disclosures of Former National Security Agency Contractor Edward Snowden*, September 15, 2016, https://intelligence.house.gov/uploadedfiles/hpsci_snowden_review_declassified.pdf; US House of Representatives Permanent Committee on Intelligence, Press Release, September 15, 2016, https://intelligence.house.gov/news/documentsingle.aspx?DocumentID=692.

12. Barton Gellman, "Edward Snowden, after Months of NSA Revelations, Says His Mission's Accomplished," *Washington Post*, December 23, 2013; Andrea Peterson, "Snowden: I Raised NSA Concerns Internally over 10 Times before Going Rogue," *Washington Post*, March 7, 2014.

6. ARE ROTTEN APPLES OR ROTTEN BARRELS RESPONSIBLE FOR TECHNOLOGICAL MISHAPS?

1. Diane Vaughan, *The Challenger Launch Decision: Risky Technology, Culture, and Deviance at NASA* (Chicago: University of Chicago Press, 1996).

2. "Interview: Diane Vaughan," ConsultingNewsLine, Quantorg, May 2008, http://www.consultingnewsline.com/Info/Vie%20du%20Conseil/Le%20Consultant%20du%20mois/Diane%20Vaughan%20%28English%29.html. Used with permission by Diane Vaughan.

3. See "Fukushima Daiichi Accident," World Nuclear Association, updated October 2018, http://www.world-nuclear.org/information-library/safety-and-security/safety-of-plants/fukushima-accident.aspx.

4. United States, Presidential Commission on the Space Shuttle Challenger Accident, *Report to the President: Actions to Implement the Recommendations of the Presidential Commission on the Space Shuttle Challenger Accident* (National Aeronautics and Space Administration, July 14, 1986), https://history.nasa.gov/rogersrep/genindex.htm.

5. R. D. Marshall et al., *Investigation of the Kansas City Hyatt Regency Walkways Collapse* (Washington, DC: National Bureau of Standards, 1982), https://doi.org/10.6028/Nbs.bss.143.

6. Sarah K. A. Pfatteicher, " 'The Hyatt Horror': Failure and Responsibility in American Engineering," *Journal of Performance of Constructed Facilities* 14, no. 2 (2000): 63.

7. Gregory P. Luth, "Chronology and Context of the Hyatt Regency Collapse," *Journal of Performance of Constructed Facilities* 14, no. 2 (2000): 51–61; William Robbins, "Engineers Are Held at Fault in '81 Hotel Disaster, *New York Times*, November 16, 1985.

8. Pffateicher, " 'Hyatt Horror,' " 64.

9. Robbins, "Engineers Are Held at Fault."

10. Luth, "Chronology and Context of the Hyatt Regency Collapse," 51.

11. Report of the Columbia Accident Investigation Board, *Space Shuttle Columbia and Her Crew* (NASA, August 26, 2003), https://www.nasa.gov/columbia/home/CAIB_Vol1.html.

12. "NASA's 'Broken Safety Culture,' " *Washington Post*, August 27, 2003.

13. Anton R. Valukas, *Report to Board of Directors of General Motors Company Regarding Ignition Switch Recalls* (Jenner & Block, Detroit, May 29, 2014), 1, http://www.beasleyallen.com/webfiles/valukas-report-on-gm-redacted.pdf.

14. Bill Vlasic, "G.M. Begins Prevailing in Lawsuits over Faulty Ignition Switches," *New York Times*, April 10, 2016.

15. Valukas, "Report," 251.

16. Ibid., 255.

17. Ibid., 3.

18. Ibid., 100–101.

19. Prohibited by the Clean Air Act, a defeat device is "any device that bypasses, defeats, or renders inoperative a required element of the vehicle's emissions control system." See https://www.epa.gov/vw/laws-and-regulations-related-volkswagen-violations.

20. Russell Hotten, "Volkswagen: The Scandal Explained," *BBC News*, December 10, 2015, https://www.bbc.com/news/business-34324772.

21. Jan Schwartz and Victoria Bryan, "VW Dieselgate Bill Hits $30 Bln after Another Charge," *Reuters*, September 29, 2017, https://www.reuters.com/article/legal-uk-volkswagen-emissions/vws-dieselgate-bill-hits-30-bln-after-another-charge-idUSKCN1C4271.

22. Geoffrey Smith and Roger Parloff, "Hoaxwagen," *Fortune*, March 7, 2016, http://fortune.com/inside-volkswagen-emissions-scandal/.

23. For an estimate of the human health costs, see R. Oldenkamp, R. van Zelm, and M. A. Huijbregts, "Valuing the Human Health Damage Caused by the Fraud of Volkswagen," *Environmental Pollution* 212 (2016): 121–27.

24. David Shepardson and Edward Taylor, "Ex-Volkswagen CEO Winterkorn Charged in U.S. over Diesel Scandal," *Reuters*, May 3, 2018, https://www.reuters.com/article/us-volkswagen-emissions/ex-volkswagen-ceo-winterkorn-charged-in-u-s-over-diesel-scandal-idUSKBN1I42I3.

25. Prachi Patel, "Engineers, Ethics, and the VW Scandal," *IEEE Spectrum*, September 25, 2015, http://spectrum.ieee.org/cars-that-think/at-work/education/vw-scandal-shocking-but-not-surprising-ethicists-say.

26. Department of Justice Office of Public Affairs, "Volkswagen Engineer Sentenced for His Role in Conspiracy to Cheat U.S. Emissions

Tests," August 25, 2017, https://www.justice.gov/opa/pr/volkswagen-engi
neer-sentenced-his-role-conspiracy-cheat-us-emissions-tests.

27. Andreas Cremer and Tom Bergin, "Fear and Respect: VW's Culture
under Winterkorn," *Reuters,* October 10, 2015, http://www.reuters.com
/article/us-volkswagen-emissions-culture-idUSKCN0S40MT20151010.

28. Jack Ewing and Graham Bowley, "The Engineering of Volkswagen's Aggressive Ambition," *New York Times,* December 13, 2015.

29. Leah McGrath Goodman, "Why Volkswagen Cheated," *Newsweek,*
December 15, 2015, http://www.newsweek.com/2015/12/25/why-volkswagen-
cheated-404891.html.

30. J. S. Nelson, "The Criminal Bug: Volkswagen's Middle Management" (April 19, 2016), 18, available at SSRN: https://papers.ssrn.com
/sol3/papers.cfm?abstract_id=2767255.

31. Ewing and Bowley, "Engineering of Volkswagen's Aggressive
Ambition."

32. Smith and Parloff, "Hoaxwagen."

7. WILL AUTONOMOUS CARS EVER BE SAFE ENOUGH?

1. See the NHTSA website at https://www.nhtsa.gov/technology
-innovation/automated-vehicles-safety.

2. Jeffrey K. Gurney, "Sue My Car Not Me: Products Liability and
Accidents Involving Autonomous Vehicles," *University of Illinois Journal
of Law, Technology and Policy* (2013): 251. Footnotes have been removed
from this quotation.

3. Gary Silberg et al., "Self-Driving Cars: The Next Revolution"
(KPMG and Center of Automotive Research, 2012), http://www.kpmg-
institutes.com/institutes/manufacturing-institute/articles/2017/11/self-
driving-cars-the-next-revolution.html.

4. Ibid.

5. M. L. Cummings and J. Ryan, "Point of View: Who Is in Charge?
The Promises and Pitfalls of Driverless Cars," *TR News* 292 (2014): 292.

6. M. L. Cummings, "Hands Off: The Future of Self-Driving Cars,"
Testimony before the US Senate, March 15, 2016, https://governmentrela
tions.duke.edu/wp-content/uploads/Cummings-Senate-testimony-2016.
pdf.

7. Timothy B. Lee, "Self-Driving Cars Are a Privacy Nightmare; and
It's Totally Worth It," *Washington Post,* May 21, 2013.

8. The Tesla Team, "A Tragic Loss," Tesla.com, June 30, 2016, https://
www.tesla.com/blog/tragic-loss.

9. Aarian Marshall and Alex Davies, "Uber's Self-Driving Car Saw
the Woman It Killed, Report Says," *Wired,* May 24, 2018, https://www
.wired.com/story/uber-self-driving-crash-arizona-ntsb-report/.

10. Christina Bonnington, "Lack of Focus," *Slate,* March 26, 2018, https://slate.com/technology/2018/03/safety-drivers-attention-spans-might-slow-self-driving-car-progress.html.

11. Jay Bennett, "Self-Driving Cars Could Cause a Major Organ Shortage," *Popular Mechanics,* December 30, 2016, https://www.popularme chanics.com/cars/car-technology/a24570/self-driving-cars-could-cause-massive-organ-shortage/.

12. Alex Davies, "Self-Driving Cars May End the Fines That Fill City Coffers," *Wired,* July 14, 2015, https://www.wired.com/2015/07/self-driving-cars-may-end-fines-fill-city-coffers/.

13. Jack Boeglin, "The Costs of Self-Driving Cars: Reconciling Freedom and Privacy with Tort Liability in Autonomous Vehicle Regulation," *Yale Journal of Law and Technology* 17 (2015): 171.

14. Jason Borenstein, Joseph R. Herkert, and Keith W. Miller, "Self-Driving Cars and Engineering Ethics: The Need for a System Level Analysis," *Science and Engineering Ethics* (in press), https://doi.org/10.1007/s11948-017-0006-0.

15. Keith Naughton, "U.S. Wants Safety 'Two Times Better' for Autonomous Cars," *Insurance Journal,* June 9, 2016, https://www.insurance journal.com/news/national/2016/06/09/411345.htm.

16. P. Liu, R. Yang, and Z. Xu, "How Safe Is Safe Enough for Self-Driving Vehicles?," *Risk Analysis* 39, no. 2 (2019): 315–25.

17. Nidhi Kalra and Susan M. Paddock, "Driving to Safety: How Many Miles of Driving Would It Take to Demonstrate Autonomous Vehicle Reliability?" (Rand Corporation Research Report, RR-1478-RC, 2016), https://doi.org/10.7249/RR1478.

18. M. L. Cummings, "The Brave New World of Driverless Cars," *TR News* 308 (2017): 34.

19. Cummings and Ryan, "Point of View."

20. Cummings, "Brave New World."

21. See, e.g., https://autonomousweapons.org.

8. IS SOCIAL JUSTICE IN THE SCOPE OF ENGINEERS' SOCIAL RESPONSIBILITIES?

1. Sylvia H. Washington and David Pellow, "Water Crisis in Flint, Michigan: Interview with David Pellow, Ph.D.," *Environmental Justice* 9, no. 2 (2016): 54.

2. The list is available at www.greatachievements.org.

3. See, e.g., Donna Riley, *Engineering and Social Justice* (San Rafael, CA: Morgan and Claypool, 2008).

4. The phrase "dumping in Dixie" originated with the work of Robert Bullard. See Robert D. Bullard, *Dumping in Dixie: Race, Class, and Environmental Quality* (Boulder, CO: Westview, 2008).

5. For evidence of environmental injustice, see, e.g., Gordon Walker, *Environmental Justice, Concepts, Evidence and Politics* (London: Routledge, 2012).

6. See https://www.un.org/millenniumgoals/; and https://unfounda tion.org/what-we-do/issues/.

7. Riley, *Engineering and Social Justice*, 4.

8. Langdon Winner, "Do Artifacts Have Politics?," *Daedalus* 109, no. 1 (1980): 121–36.

9. Robert Moses was not an engineer but a powerful figure in New York government, the powerhouse behind many public projects in the state during the early twentieth century.

10. Daniel McGlynn, "Engineering Social Justice," *Berkeley Engineer*, Spring 2014, https://engineering.berkeley.edu/magazine/spring-2014/en gineering-social-justice.

11. See the website of Engineers without Borders US, https://www .ewb-usa.org.

12. Rachel N. Weber, "Manufacturing Gender in Commercial and Military Cockpit Design," *Science, Technology, and Human Values* 22, no. 2 (1997): 235–53.

13. Colbe Itkowitz, "The Heroic Professor Who Helped Uncover the Flint Lead Water Crisis Has Been Asked to Fix It," *Washington Post*, January 27, 2016. Marc Edwards received the 2018 AAAS Scientific Freedom and Responsibility Award for his activities in documenting and drawing attention to the dangerous levels of lead contamination in Flint. Nevertheless, I must note that a few years after working closely with Flint residents, Edwards came into conflict with some of the activists who were still working on the water contamination problem. The real lives of engineers are complicated!

INDEX

AAA. *See* American Automobile
Association
Abaté, Charles, 19, 21
ABET, 2, 10, 31–32, 34,
181n1(ch3)
accidents and disasters: anticipating and building for, 120;
automobile accidents and
fatalities, 106, 124, 128, 141,
143–44, 145, 146; cultural/
environmental causes
("rotten barrels"), 117–19,
126–35 (*see also* NASA);
ethical design issues of
accident avoidance systems,
152–54; individual actions ("rotten apples") and, 117–18,
123–26, 130–31, 133–35;
responsibility, causality, blame,
accountability, and liability, 6,
119–23, 125–26, 128–29. *See
also* safety; *and specific
incidents*
accountability: for accidents,
mishaps, or disasters, 122–23,
125–26, 128–29; for intentional
wrongdoing, 132. *See also*
accidents and disasters; *and
specific incidents*

act utilitarianism, 56. *See also*
utilitarianism
advanced driver assistance
systems (ADAS), 139–40
African Americans, 157, 167. *See
also* minority groups
airbags, 128, 130. *See also* General
Motors (GM) ignition switch
case
airplane design, 173
American Automobile Association
(AAA), 141
American Institute of Mining and
Metallurgical Engineers
(AIMME), 1, 32
American Society of Civil
Engineers (ASCE), 1, 26, 28, 32,
38, 182n2
American Society of Mechanical
Engineers (ASME), 1, 28, 32,
43, 182n4
arete (virtues), 64–65. *See also*
virtue ethics
Aristotle, 63, 64
artifacts, politics of, 167–68
ASCE. *See* American Society of
Civil Engineers
ASME. *See* American Society of
Mechanical Engineers

automobiles: accident and death statistics, 141, 145; advanced driver assistance systems (ADAS), 139–40; defeat devices, 131, 133, 186n19; Ford Pinto design flaw, 105–6, 107–8, 117, 146; General Motors ignition switch case, 124, 127–31; infant car seat design, 67; meaning of, 146–47, 150–51; safety systems and components, 148, 155; as sociotechnical systems, 148–49; Volkswagen emissions fraud, 124, 131–34. *See also* autonomous cars

autonomous cars, 137–55; "autonomous" as term, 150; as emerging technology, 138–39, 147–48; ethical design issues of accident avoidance systems, 62–63, 152–54; fatalities involving, 143–44; innovations applied to human-driven cars, 155; Kantian theory and, 63; level of safety (acceptable risk), 143; levels of automation and terminology, 139–40, 144, 151; operator(s), 139–40, 144, 149; pros and cons, 140–43; public acceptance, 150; safety and cost-benefit analysis, 144–48; safety and risk in sociotechnical system of, 148–51; safety in mixed system, 138, 151; safety potential, 141–42, 144, 145, 154; safety through standards and regulation, 137, 151–52; and social justice, 147, 158; uncertainties about, 138–39, 147; utilitarianism and, 62–63, 147, 153; vulnerable to hackers, 142

autonomous weapons, 154

autonomy: adoption of codes of ethics and, 38; collective autonomy, 31–32; limited autonomy of engineers, 169–70; as mark of professions, 31, 33, 34; public trust and, 37; respecting each person's autonomy, 61; technical autonomy of engineers, 33, 34

Barra, Mary, 129

BBC News, 90

Bechtel Corporation, 94, 98–99

Bentham, Jeremy, 53. *See also* utilitarianism

BER. *See* Board of Ethical Review

bid rigging, 89, 91. *See also* corruption

biomedical devices, 49, 51, 168

blame, 123. *See also* accidents and disasters

Bloom, Paul, 13

Board of Ethical Review (BER). *See under* National Society of Professional Engineers

Bowley, Graham, 134

bribery, 88–89; gift vs. bribe, 3–4; Mulroney case, 9; Odebrecht case, 90; possible consequences, 91; prohibitions on, in codes of ethics, 42, 43; reporting, 98. *See also* corruption

bridges. *See* transportation infrastructure

British Royal Academy of Engineering, 26

building construction and design, 120, 123–26

cars. *See* automobiles

Castro, Salvador, 93–94, 99

categorical imperative, 60–62, 80, 88, 103, 162

cell phones, 49, 149–50

Challenger space shuttle disaster, 117, 118, 122–23, 127

Citicorp Headquarters, 23

Clean Air Act, 131, 186n19

clients, relationships with: categorical imperative (Kantian theory) and, 61; clients as employers, 73; codes of ethics on engineers' obligations, 27,

28, 34–35; confidentiality, 74, 78–80; conflicts of interest, 73–74, 83–88; corruption, 88–91; engineers' conflicting responsibilities, 73–74, 91; guns-for-hire view, 74–75, 77, 79–80, 83, 170; honesty, 80–83; obligation to client vs. public safety/welfare, 18–19, 29, 75–78; virtue theory and, 65. *See also* whistleblowers and whistle-blowing

codes of ethics (engineering): applicability, 43; collective wisdom and development process, 41, 45; and conflict of interest, 87; and corruption, 90–91; criticisms, 26, 29–30, 37, 39–40, 44; effectiveness, 26, 44; and employer expectations, 38, 41, 44; enforcement, 29–30, 36, 39, 40, 44; engineering as profession and, 30–35, 44; engineering ethics codified in, 3; fundamental principles, 27–29; guns-for-hire view and, 75; interpreting and applying, 18–19; on loyalty toward employers, 100; multiple audiences, 38–39, 41, 43; multiple codes vs. single global, unifying code, 43–44; need for, 4, 26, 30, 36–42, 44–45; new engineers guided by, 41, 45; normative orientation, 34; paramountcy clause, 19, 27, 28, 34–35, 42, 75, 80, 143, 158–59 (*see also* public safety); as principles rather than rules, 39–41; and public trust, 37–38; reinforcement activities, 37; similarities across fields, 27–28, 44; virtue ethics and, 65–66; what should be included, 42–43. *See also specific organizations*

codes of ethics (generally), 26

codes of professional conduct (generally), 14–19

coercion, 61

collective autonomy, 31

Columbia space shuttle disaster, 124, 126–27

commitment to service, 33, 34–35. *See also* clients, relationships with; public safety

communications networks, 49

compensatory justice, 164

competence in engineering, 14. *See also* expert knowledge; undergraduate engineering programs

complacency criticism, 39–40

computer scientists, 159

computer software, 171, 173; of autonomous cars, 62–63, 142, 153–54; conflict of interest in evaluating, 85–86; defeat devices, 131, 133 (*see also* Volkswagen emissions fraud); for Facebook, 50

confidentiality, 74, 78–80

conflicts of interest, 15–16, 73–74, 83–88. *See also* clients, relation-ships with

consequentialism, 54. *See also* utilitarianism

construction: building construc-tion and design, 120, 123–26; corruption in the construction industry, 89–90

ConstructionDive, 89

contracts (employment con-tracts), 101

corruption, 88–91, 169. *See also* bribery

cost-benefit analysis, 55, 144–48

courage, moral, 23–24. *See also* whistleblowers and whistle-blowing

Cummings, Mary, 142, 152

debate format, 2–5, 19, 178

decision making. *See* ethical reasoning; practical ethics; theoretical ethics

Defense Nuclear Facilities Safety Board, 94, 98

De George, Richard, 105–9, 111

deontological theories, 59. *See also* Kantian theory

Department of Energy, 94

design: design problems, 66–68, 120–21, 123, 124–26 (see also *Challenger* space shuttle disaster); and diversity and social justice, 167–69, 173, 175

disabilities, people with, 141, 173

disasters. *See* accidents and disasters

disclosure, 86–87, 101

discrimination: by employers, 101, 102; Flint water crisis and, 156–57; prohibited under codes of ethics, 42; through design, 167, 168–69. *See also* minority groups; social justice

distributive justice, 163–64. *See also* social justice

diversity, 173, 175. *See also* disabilities, people with; minority groups; poverty; women

driverless cars. *See* autonomous cars

"dumping in Dixie" phenomenon, 162, 188n4

education in engineering. *See* undergraduate engineering programs

Edwards, Marc, 176, 189n13

egoism, 55

Elegido, Juan, 102

embezzlement, 88. *See also* corruption

employers, relationships with: blind obedience vs. moral responsibility, 51–52 (*see also* guns-for-hire position); clients as employers, 73; codes of ethics and employer expectations, 38–39, 41, 44; codes of ethics on engineers' obligations, 27, 28; conflicts of interest, 15–16, 74; contractual nature of, 101; doctrine of employment at will, 101–2, 103; engineers' autonomy and decision-making power, 169–70; ethical implications of choice of employer, 172–73; as instrumental relationships, 102–3, 104; laws governing, 101–2; loyalty, 96, 100–105, 107; managers' vs. engineers' roles, 107, 108; professional standards and, 32–33, 38, 93; reporting safety concerns, 29, 32–33, 66, 80, 93–94, 105–6, 107–9; whistleblowers as heroes or traitors, 95–96, 113–14, 178. *See also* clients, relationships with; whistleblowers and whistleblowing

employment at will doctrine, 101–2, 103

enduring questions, 2, 177

enforcement of codes of ethics, 29–30, 36, 39, 40, 44

engineering (field). *See* profession of engineering

engineering ethics: challenge of determining right action(s), 3; engineering as inherently ethical endeavor, 50, 179; growing interest in, 1–2; identifying ethical issues, 19–21, 24; as professional ethics, 46, 47–53, 68–69. *See also* codes of ethics (engineering); ethical reasoning; ethics (generally); moral values; practical ethics; teachability of engineering ethics; theoretical ethics; *and specific topics*

Engineers without Borders (EWB), 172

environmental harm: fictional example, 81–83; Flint, Michigan, water crisis, 156–57, 162–63, 176, 189n13; Volkswa-

gen emissions fraud, 124,
131–34. *See also* environmental
injustice; Hanford Waste
Treatment Plant
environmental injustice: defined,
157; Flint, Michigan, water
crisis, 156–57, 162–63, 176,
189n13; human rights and,
164–65; intergenerational,
159–60; types of justice and,
163–64; utilitarianism and,
161–62. *See also* environmental
harm
Environmental Protection Agency
(EPA), 131
EPA. *See* Environmental Protection Agency
equality and inequality. *See*
environmental injustice;
fairness; minority groups;
poverty; social justice
ergon (function), 63. *See also*
virtue ethics
Ericksson, Stefan, 18–19
Ethical Problems in Engineering
(Alger et al.), 181
ethical reasoning: ethics education and, 11; Kantian theory on,
59–60; and moral intuitions,
11–13; ongoing/iterative nature
of decision-making, 66–67,
76–77, 82; practical ethical
decision-making (real-world
situations), 47, 66–68, 69;
teaching ethical decision-making, 22; trolley problem,
152–53; virtue ethics and,
65–66. *See also* Kantian theory;
utilitarianism
ethics (generally): codes of ethics,
26; fairness at heart of, 160;
motivations for ethical/moral
behavior, 11–13; theoretical
ethics, 46–47, 53–66. *See also*
engineering ethics; ethical
reasoning; Kantian theory;
moral values; practical ethics;
utilitarianism; virtue ethics

eudaimonia, 64–65. *See also* virtue
ethics
EWB. *See* Engineers without
Borders
Ewing, Jack, 134
expert knowledge (expertise), 14,
33–34, 37; guns-for-hire view
and, 75, 77; and whistleblowing, 105–6, 107–8. *See also*
knowledge
extortion, 61/

Facebook, 49–50
fairness: autonomous cars and,
147; of codes of ethics, 43;
concepts of justice and, 160,
162–65; intergenerational
injustice, 159–60; in law and
sports, 83–84, 86. *See also* social
justice
financial interests (personal), and
conflict of interest, 85–86
financial problems (organizational), and safety, 129
Flint, Michigan, water crisis,
156–57, 162–63, 176, 189n13
Ford Pinto design flaw, 105–6,
107–8, 117, 146
foundational issues in engineering ethics, 5, 177. *See also*
specific issues
freedom, as value, 146–47
friendship, 13, 20–21, 54, 84,
100–101
Fukushima Daiichi Nuclear
Power Plant disaster, 119–22
fundamental principles, of codes
of ethics, 27–29

gender, and design, 173
General Motors (GM) ignition
switch case, 124, 127–31
Global Infrastructure Anti-Cor-
ruption Centre, 90
GM. *See* General Motors (GM)
ignition switch case
Goodman, Leah McGrath, 133
Google, 141

Gorman, Elizabeth, 33, 34–35
"Grand Challenges" for engineering (NAE), 171
Greatest Engineering Achievements of the 20th Century (list), 159
guns, 168. *See also* weapons
guns-for-hire position: engineers as professionals vs. as hired guns, 74–75, 91, 170; problems with guns-for-hire view, 75, 77, 79–80, 83, 170. *See also* clients, relationships with; employers, relationships with
Gurney, Jeffrey, 140–41

Haidt, Jonathan, 12, 13
Hanford Waste Treatment Plant, 94, 98–99
happiness: *eudaimonia*, 64 (*see also* utilitarianism); as highest good, 54–55, 57–58, 59
Harris, Charles, 43–44, 65–66
Helgesson, Gert, 18–19
Höglund, Anna, 18–19
honesty: complexities, 40; and engineer-client relationships, 80–83; in Kantian theory, 59, 61; rule utilitarianism and, 56; in various codes of ethics, 27, 40; as virtue, 64. *See also* integrity
hotel collapse, 123–26
House of Representatives Permanent Select Committee on Intelligence, 110
human life, value of, 146. *See also* public safety; safety
human reasoning, 59–60
human rights, 164–65
Hunter Roberts Construction Group, 89–90
Hyatt Regency Hotel collapse, 123–26

IEEE. *See* Institute of Electrical and Electronics Engineers

illegal behavior, 42. *See also* bribery; corruption; Volkswagen emissions fraud
impartiality, 83–84, 86. *See also* conflicts of interest; fairness
income, 33, 35
infant car seat design, 67
informed consent, 152
infrastructure. *See* transportation infrastructure
inspiration, 23–24
Institute of Electrical and Electronics Engineers (IEEE): Code of Ethics, 28, 40, 87, 182n3; Salvador Castro case, 93–94
instrumental goods, 54
integrity: of engineers and the engineering profession, 27, 28, 75, 81–82, 88; Kantian theory on, 61–62. *See also* corruption; honesty; impartiality
Internet, 166, 179
intrinsic goods, 54–55. *See also* happiness
intuition, moral, 11–13

Japanese earthquake and tsunami, 119–20. *See also* Fukushima Daiichi Nuclear Power Plant disaster
Jones, Sharon, 42
judges, 83–84, 86. *See also* law
judgment, professional, and conflict of interest, 85–88. *See also* conflicts of interest
justice, concepts of, 160–65. *See also* social justice

Kadir, Khalid, 168
Kalra, Nidhi, 151–52
Kansas City Hyatt Regency Hotel collapse, 123–26
Kant, Immanuel, 59–62. *See also* Kantian theory
Kantian theory, 53, 58, 59–62; and autonomous cars, 63; categorical imperative, 60–62, 88, 103,

162; critiques, 62; and social justice, 162–63; and the trolley problem, 153. *See also* categorical imperative
knowledge: expert knowledge (technical expertise), 14, 33–34, 37, 75, 77; and happiness, 55; knowledge sharing and safety issues, 130; of professional norms, practices, and codes, 14–19; technical expertise and whistleblowing, 105–6, 107–8

law: as concept in Kantian theory, 59–60; judges' impartiality/fairness, 83–84, 86; training and admittance to profession, 30–31
lead poisoning, 156–57. *See also* Flint, Michigan, water crisis
legalistic thinking, 19
LeMessurier, William, 23
liability, 123. *See also* accidents and disasters
Liang, James, 133
licensing, 29, 35–36, 39, 49
Long Island, New York, 167
loyalty, 96, 98, 100–105, 107. *See also* whistleblowers and whistleblowing
Luth, Gregory, 126
lying. *See* honesty

macro-ethical issues, 5, 6, 178. *See also* autonomous cars; social justice; *and other topics*
Martin, Mike, 38, 96–97, 99
maxims (Kantian concept), 61–62
medicine (field): biomedical devices, 49, 51, 168; biomedical engineers, 168; doctors as employees, 33; training and admittance to, 30–31
Metropolitan Transit Authority (MTA), 89

Michelfelder, Diane, 42
Michigan: Department of Environmental Quality, 156–57. *See also* Flint, Michigan, water crisis
micro-ethical issues, 5, 178. *See also specific topics*
Mill, John Stuart, 4, 53. *See also* utilitarianism
minority groups, 156–57, 161–62, 167, 174–75. *See also* poverty; social justice
money, 54. *See also* bribery; corruption
moral values: early acquisition, 9–10; familiarity with moral concepts/frameworks, 21; Kantian theory and, 58, 59–62, 63; moral courage, 23–24; moral intuitions and ethical reasoning, 11–13, 57; morality defined, 60; motivations for moral behavior, 11–13, 22–24; teaching moral reasoning and decision-making, 22; utilitarianism and, 53–58, 59, 62–63; virtue ethics, 53, 63–66. *See also* ethical reasoning; ethics (generally); teachability of engineering ethics; theoretical ethics; whistleblowers and whistleblowing
Moses, Robert, 167–68, 189n9

NAE. *See* National Academy of Engineering
NASA, 117, 118, 122–23, 124, 126–27
National Academy of Engineering (NAE), 159, 171
National Bureau of Standards, 124
National Council of Examiners for Engineering and Surveying, 36
National Highway and Traffic Safety Association (NHTSA), 139, 152

National Security Agency (NSA), 95, 110–13. *See also* Snowden, Edward

National Society of Professional Engineers (NSPE): Board of Ethical Review, 14–17, 75–80, 87; Code of Ethics, 15, 16, 18, 26, 27, 40, 87, 182n1; licensure requirements for professional engineers, 35–36

negligence. *See* Kansas City Hyatt Regency Hotel collapse

Nelson, J. S., 133

Newsweek, 133

New York Times, 89, 110

NHTSA. *See* National Highway and Traffic Safety Association

Nicomachean Ethics (Aristotle), 63

norms: knowledge of professional norms, practices, and codes, 14–19; normative orientation of codes of ethics, 3, 34; and technical mishaps, 120–22

NSA. *See* National Security Agency

NSPE. *See* National Society of Professional Engineers

Nunes, Devin, 110

objectivity, 15–16, 84, 86. *See also* conflicts of interest; impartiality

Occupational Safety and Health Administration (OSHA), 94

Odebrecht corporation, 90

ombudsmen, 114

organizational cultures (environments): and accidents, disasters, and wrongdoing, 117–19, 123–24, 126–34; freedom to express concerns, 114, 134

OSHA. *See* Occupational Safety and Health Administration

Paddock, Susan, 151–52

Parloff, Roger, 134

Partnering against Corruption Initiative, 90

payment, promptness of, 61–62

Pellow, David, 157

personal life, 20–21, 84–85, 87, 100–101. *See also* friendship

PEs. *See* professional engineers

Pfatteicher, Sarah, 38, 124–26

Piëch, Ferdinand, 134. *See also* Volkswagen emissions fraud

Pinto. *See* Ford Pinto design flaw

Pizzaro, David, 13

policy making, 55

politics of artifacts, 167–68

poverty: and corruption, 89; and human rights, 165; poor/low-income people as engineers, 175; and social/environmental injustice, 156–57, 161–62, 167. *See also* social justice

practical ethics, 47, 69; as design problem (multiple solutions), 66–68, 76–77, 82; virtue theory and, 65. *See also* ethical reasoning; *and specific issues and cases*

Pritchard, Michael, 23

privacy and privacy rights, 95, 142, 146–47. *See also* National Security Agency

procedural justice, 163–64. *See also* social justice

professional engineers (PEs), 35–36, 39

professional ethics, engineering ethics as, 46, 47–53, 68–69

professional societies (engineering organizations), 10, 39, 43, 49. *See also* codes of ethics (engineering); *and specific organizations*

profession of engineering: autonomy of engineers, 31–32, 33, 34, 37, 38, 169–70; in business context, 91; characteristics of professions, 33–35; and confidentiality, 79–80; conflict of interest and, 86–87, 88; corruption and, 88; engineering as profession, 30–35, 44, 91; engineering ethics as professional ethics, 46, 47–53, 68–69, 91; engineers as professionals vs. engineers as employees or hired guns, 32–33, 74–75, 91, 170; expert knowledge, history of, 1; learning professional behavior, 41; licensing, 29, 35–36, 39, 49; need for codes of ethics, 4, 26, 30, 36–42, 44–45; number of practicing engineers, 36; professional engineers, 35–36; public trust and, 36–38, 43, 82, 86, 88; semi-regulatory responsibilities, 133; social responsibility of engineers, 46, 50–52, 157–60, 165–72, 175–76; training; virtues of engineering, 65–66; women and minorities in, 174–75. *See also* employers, relationships with

professions (generally), 31, 33. *See also* law; medicine; profession of engineering

promise keeping, 61–62. *See also* integrity

public safety (health, welfare): automobile fatalities (statistics), 141, 145; vs. engineers' obligations to client or employer, 18–19, 29, 80, 93; fire-alarm case (example), 75–78; Flint, Michigan, water crisis, 156–57, 162–63, 176, 189n13; Ford Pinto design flaw, 105–6, 107–8, 117, 146; General Motors ignition switch case, 124, 127–31; Kansas City Hyatt Regency Hotel collapse, 123–26; paramountcy of, 19, 27, 28, 34–35, 42, 75, 76, 80–81, 143, 158–59; and whistleblowing, 108–9. *See also* accidents and disasters; autonomous cars; environmental harm; environmental injustice; human rights; safety; social justice; whistleblowers and whistleblowing

public trust: Flint water crisis and, 163; profession of engineering and, 36–38, 43, 82, 86, 88. *See also* trust

racism. *See* Flint, Michigan, water crisis

Reagan, Ronald, 122–23

reasoning. *See* ethical reasoning; human reasoning

relationships. *See* clients, relationships with; employers, relationships with; friendship

reputation: confidentiality and, 79; of engineering profession as a whole, 27, 37, 38, 41, 86; honesty and, 81–83; whistleblowing and organization reputation, 94–95. *See also* public trust

responsibility: for accidents, mishaps, and disasters, 6, 122–23; of engineers. *See* clients, relationships with; codes of ethics; employers, relationships with; profession of engineering; social justice

Reuters, 89

Riley, Donna, 165

risk: acceptable levels, 143, 177; engineers' responsibility to report, 52, 76–77; of ethical behavior, on career, 23, 93–95, 98; fair/unfair distribution of, 156–57, 158, 161–64, 165; informed consent, 152; organizational culture and the normalization of, 118–19; role in cost-benefit analysis, 55, 144–45; in sociotechnical systems, 148–51; of whistleblowing, 93–95, 96, 98, 109, 112. *See also* safety

roads. *See* transportation infrastructure

"rotten apples and rotten barrels."
See accidents and disasters;
corruption; organizational
cultures; wrongdoing
rule utilitarianism, 55–57, 79–80.
See also utilitarianism
Ryan, Jason, 142, 152

SAE. See Society of Automotive
Engineers
safety, 6, 177; automobile death
statistics, 141, 145; cost-benefit
analysis, 144–45; reporting
safety concerns, 29, 32–33, 66,
68, 80, 93–94, 105–6, 107–9;
and risk in sociotechnical
systems, 148–51. See also
autonomous cars; public safety;
risk
Sandefur, Rebecca, 33, 34–35
Schinzinger, Roland, 38, 96–97,
99
self-driving cars. See autonomous
cars
service, commitment to, 33,
34–35. See also public safety
slavery, 57–58, 60
Smith, Geoffrey, 134
Snowden, Edward, 95–96, 105,
110–13
social intuitionist model, 11–12
social justice, 156–76; autonomous cars, unequal impacts of,
147, 158; compensatory justice
and, 164; decision-makers,
169–70; design and, 167–69,
172–73; distributive vs.
procedural justice and, 163–64;
engineering's relationship to, 6;
Flint water crisis as environmental injustice, 156–57,
162–63, 176, 189n13; good of
the many vs. rights of the few,
57–58; human rights and,
164–65; intergenerational
injustice, 159–60; Kantian
theory and, 162–63; social
responsibility of engineers, 46,

50–53, 157–60, 165–72, 175–76;
and sustainability, 42–43, 165;
as term, 165; transportation
infrastructure and, 167;
utilitarianism and, 57–58, 147,
160–63
social status, 33, 35
society and engineering: engineering, society, and technology
intertwined, 48–50, 166–69,
170–72, 179; social responsibility of engineers, 46, 50–53,
157–60, 165–72, 175–76. See
also technology; social justice
Society of Automotive Engineers
(SAE), 139–40
Spiegel, Der, 133
sports, 84, 86
state licensing boards, 29, 36, 39.
See also licensing
Stephan, Karl, 9–10
sustainability, 28, 42–43, 165

Tamosaitis, Walter, 94, 98–99
teachability of engineering ethics,
6, 9–25; cognitive component
(identifying ethical issues),
19–21, 24; core moral values,
9–10; debate format, 3–5, 19,
178; early acquisition of morals
vs., 9–10; ethical/moral
reasoning, 11–13, 22, 24; goals
of teaching engineering ethics,
10, 17–24; improving ethical
reasoning, 11; knowledge
component, 14–17, 24; motivational component, 22–24;
skepticism about, 9–10, 11–12;
undergraduate engineering
program texts and requirements, 2, 10, 182n11
technical autonomy, 33, 34. See
also autonomy
technical knowledge/expertise.
See knowledge
technology: emerging technologies, 138–39, 147–48; engineering, society, and technology

intertwined, 48–50, 166–69,
170–72, 179 (*see also* social
justice); role and influence, 1;
safety and risk in sociotechnical
systems, 148–51; and social and
generational inequities, 159–60,
167–69; technological mishaps
(*see* accidents and disasters); as
value neutral, 167–68. *See also*
autonomous cars
telos (purpose/function), 63, 64.
See also virtue ethics
Tesla automobiles, 144
theoretical ethics, 53–66;
Kantian theory, 58, 59–62;
as means of thinking about
ethics, 46–47, 69; utilitarianism,
53–58; virtue ethics, 53,
63–66, 80, 88, 100–101. *See also*
ethical reasoning; ethics
(generally)
time sheet alteration, 89–90
Transparency International, 89
transportation infrastructure, 49,
149, 158, 167–68
trolley problem, 152–53
trust: conflicts of interest and,
86–88; engineer-client trust, 78,
81–82; public trust, 36–38, 43,
82, 88, 163. *See also* confidentiality; honesty

umpires, 84, 86
undergraduate engineering
programs: ethics training, 2, 10,
182n11; expert knowledge
acquired, 34, 37; graduates
considered engineers, 30–31
Union Carbide explosion (Bhopal,
India), 117
United Nations Millennium
Development Goals, 165
universalizability, 61–62
URS Energy & Construction, 94,
98–99
utilitarianism, 53–58; and
autonomous cars, 62–63, 147,
153; and conflicts of interest,

88; and cost-benefit analysis,
55, 145, 147; critiques, 57–58,
59, 60, 147, 160, 162; rule
utilitarianism, 55–57, 79–80;
and social justice, 57–58, 147,
160–63; and the trolley problem, 153

value of human life, 146
Valukas, Anton, and Valukas
Report, 127–31
Vaughan, Diane, 118–19
vendors, relationships with, 20–21
virtue ethics, 53, 63–66; and
confidentiality, 80; and conflicts
of interest, 88; loyalty as a
virtue, 100–101; virtues (*arete*),
64–65
Volkswagen emissions fraud, 124,
131–34

Washington Post, 111–12, 127
water. *See* Flint, Michigan, water
crisis
weapons, 154, 168
whistleblowers and whistleblowing, 93–114; defined, 96–100;
dilemma of, 13, 32–33, 48, 113;
as example of professional
ethics, 47–48; as heroes or
traitors, 95–96, 113–14, 178;
internal vs. external, 97–98,
106–7, 112–13; justifying,
105–9, 110–11; need for, 96,
114; public disclosure of
wrongdoing, 94, 95, 99, 108,
111; risks and costs, 93–95, 96,
98, 109, 112; Snowden case,
95–96, 105, 110–13; whistleblower protection laws, 103;
whistleblowing as disloyalty, 96,
98, 102, 103; whom to approach
and what to reveal, 113
Whitbeck, Caroline, 67–68
Winner, Langdon, 167–68
Winterkorn, Martin, 132, 134.
See also Volkswagen emissions
fraud

women, 165, 173, 174–75
World Economic Forum, 90
World Federation of Engineering
    Organizations, 26
World Health Organization, 141
Worthington, Peter, 9

wrongdoing. *See* accidents and
    disasters; bribery; corruption;
    Flint, Michigan, water crisis;
    Ford Pinto design flaw;
    negligence; Volkswagen
    emissions fraud